SpringerBriefs in Aging

More information about this series at http://www.springer.com/series/10048

Laetitia Teixeira · Lia Araújo ·
Constança Paúl · Oscar Ribeiro

Centenarians

A European Overview

 Springer

Laetitia Teixeira
Institute of Biomedical Sciences
Abel Salazar
University of Porto
Porto, Portugal

Constança Paúl
Institute of Biomedical Sciences
Abel Salazar
University of Porto
Porto, Portugal

Lia Araújo
Polytechnic Institute of Viseu
Viseu, Portugal

Oscar Ribeiro
Department of Education and Psychology
University of Aveiro
Aveiro, Portugal

ISSN 2211-3231   ISSN 2211-324X (electronic)
SpringerBriefs in Aging
ISBN 978-3-030-52089-2   ISBN 978-3-030-52090-8 (eBook)
https://doi.org/10.1007/978-3-030-52090-8

This Springer imprint is published by the registered company Springer Nature Switzerland AG
The registered company address is: Gewerbestrasse 11, 6330 Cham, Switzerland

Foreword

The European population is aging rapidly, what never happened in the past. Because of increasing longevity and low fertility, the share of oldest olds in the society is increasing. As a consequence of a roughly invariant age at retirement, the balance between active and non-active populations will grow in favor of the second involving numerous socio-economic implications. Nevertheless, reaching the age of 100 years or being centenarian remains exceptional and it attracts a lot of attention from both scientists and the general public. As a matter of fact, reaching 100 years is not different from 99 or 98 but our numerical system, introduced by the Romans more than two millennium ago, makes this round age a unique milestone. Living along a whole century is an exceptional situation even if nowadays it becomes more common than in the past; a French woman born in 1920 out of fifty may expect to reach 100. Most centenarians become public persons and are often officially honored by their community and the national authorities. Many European countries have a special office attached to the President or to the King/Queen that organizes ceremonies to congratulate centenarians for their 100th birthday. Recently, in some countries, their number is growing so quickly that the threshold was set at 105 years and only these semi-supercentenarians are celebrated officially.

Until recently, national statistical institutions were reluctant to publish data on age composition above age 100. Usually most statistical tables were disseminated with an open class of 80+, 90+ and sometimes 99+. Moreover, these open classes include persons with unknown age and persons whose age is not validated. Why? Because some centenarians are only administratively alive in population registers or databases whereas they probably died long time ago, somewhere abroad and their death was not officially reported. In this context, collecting and visualizing data and validated comparative figures on centenarians at European level is a challenge that Laetitia Teixeira and her colleagues succeed to face with great interest. Comparable data at European level are so much inconsistent and misleading. Accordingly, the present exercise finds its utility to describe the number and the socio-demographic characteristics of people that are part of the most increasing age group in the whole population. The unpreceded increase in number of centenarians started only a few decades ago even if centenarians were proved to exist since several centuries. Such growing trends will not slow down in the near future as the new centenarians will belong to the larger birth cohorts following WWI.

The data collection undertaken by Porto team is unique as it is the first time that comparative tables are compiled describing the main characteristics of centenarians in all European countries. Therefore, they had to face the three major problems associated to data collection: availability (and even accessibility), validity (as data on centenarians might be biased by the existence of numerous false centenarians), and comparability (a problem faced on a daily base by international statistical institutions like Eurostat). I sincerely appreciated this book as it provides for the first time some ideas on the European variability of the centenarians' main characteristics.

Michel Poulain
Emeritus Professor, UCLouvain
Ottignies-Louvain-la-Neuve, Belgium

Senior Researcher, Tallinn University
Tallinn, Estonia

Preface

As the oldest continent of the word, Europe presently faces an unprecedented demographic scenario. The number of individuals reaching very advanced ages is growing significantly, and within this age group a very particular one: centenarians.

Living up to be 100 years of age, although reachable to only a few of us, is likely to become more common, and this posits important social and health care demands. Long lives' "secrets" are to be answered by a wide range of professionals like geneticists, biologists, ecologists, and physicians, but also demographers and other social scientists that must disentangle group and individual life trajectories in historical time and space, making sense of extraordinary lives that more frequently than ever challenge our imagination as well as our capacity to deliver adequate services and friendly and inclusive societies to accommodate them.

With this book, we intend to provide a profile of European centenarians and fill a void on the available information on this population. In an eminently descriptive way, the book intends to first and foremost provide an overview of this population's characteristics in terms of sex ratio, educational level, marital status, living context and living arrangements, health profiles, and main causes of death. It does not intend to present an extensive justification on the differences observed throughout Europe nor on the geographical tendencies we might observe on their distribution and characteristics. It ultimately aims to provide researchers from these countries and from abroad who are currently working with this population, or intend to do so, an overview of their country outline in comparison with others.

The story of how this book came into being dates back to IAGG 2017 World Congress of Gerontology and Geriatrics meeting where we presented a first draft of our work on "Centenarians in Europe". Three years before we had started collecting data for the first population-based study with centenarians ever conducted in Portugal, the Oporto Centenarian Study (PT100), and wanted to have a global perspective of an up-to-date profile of exceptional longevity at a European level. In conjoining efforts to further elaborate a more detailed profile following IAGG's presentation, we started receiving a very enthusiastic feedback on the information provided. We then realized the attractiveness of the topic, particularly for researchers from other continents, but it mostly reassured the fact that limited amount of systematized information, even at a descriptive level, was available on these matters.

In conceptualizing this book, we draw on the expertise gained within the International Centenarian Consortium (ICC) that congregates researchers from all over the world and that has been having regular annual meetings since 1994. Although our presence in the ICC only dates to the meeting of 2014 in Japan, we soon realized the importance of being in touch with some of the world-reference researchers in this field. Some of the ideas we share throughout this book come from those enriching meetings and from the insightful comments we had the chance to receive.

Along with the ICC researchers, there is a wide range of people who we would like to express our gratitude. Within such a group we include all the centenarians and their caregivers we had the chance to meet and talk to, and who gave us an unique perspective on what it is like to be that old, and that surely goes far beyond what the numbers we here provide can tell us about them; furthermore, we would also like to express our gratitude to all of the PT100 research team members and faculty colleagues who kept on motivating us in writing the book.

Porto, Portugal   Laetitia Teixeira
May 2020   Lia Araújo
                                                             Constança Paúl
                                                              Oscar Ribeiro

Contents

Chapter 1
Introduction

1.1 The Oldest Old Population in the European Union

Europe is the oldest continent in the world and its population is getting remarkably older. The overall improvement of living conditions like sanitation, food access, and security, together with medical progress, lead to a substantial reduction of infant mortality and avoidance of premature death. These two outcomes were achieved in the 80s and nowadays the most common age of death is 85 years for women, and 80 years for men (Robine, 2019), which explains the raise of the oldest old population that is currently showing the highest rate of growth.

Currently, those individuals who are in their 70s, 80s or even 90s are benefiting from growing longevity. According to last statistical information on the lives of the European Union's (EU) older generations (European Union, 2019), between 2018 and 2050 the number of people aged 85 years or more in the EU is projected to more than double up 130.3%; more precisely, the number is projected to increase from 13.8 million in 2018 to 31.8 million by 2050, being that those aged 100 years or more is projected to grow from close to 106,000 in 2018 to more than half a million by 2050. Figure 1.1 shows the evolution and projections of the number of centenarians for the EU in the period 1990–2050. As we can see, the number of centenarians has been expressively growing over the last decades. It is projected to be 109,497 in 2020 and reach 509,926 by 2050. An exception could be observed for the period 2015–2010, which corresponds to the 100th anniversary of World War I.

In a closer look at the evolution and projections of the number of centenarians by EU country (Table 1.1), we can see that France, Germany and Italy are the three top countries where the number of centenarians is projected to be higher in 2050: the first with 98,317 centenarians, the second with 76,478, and the third with 74,284. Considering the percent variation of number of centenarians between 2010 and 2050 showed in Fig. 1.2, the countries where the expected increase is more expressive are Lithuania and Poland.

A complex and integrative set of individual and environmental factors, rather than a single one, explains the high number of centenarians in our societies. The century

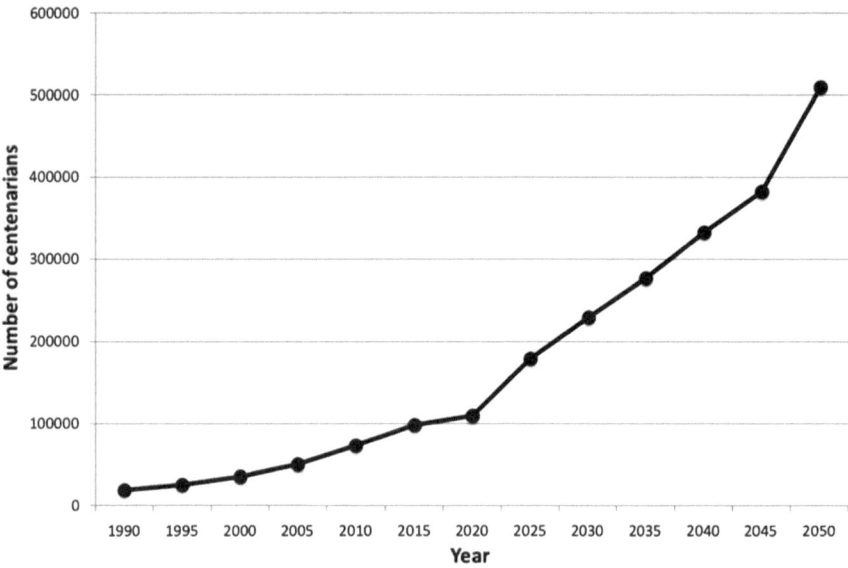

Fig. 1.1 Evolution and projections of the number of centenarians in EU (1990–2050) (*Source* UNdata | last update: 17 June 2019 | reference years: 1990 to 2050 | download date: 2 January 2020)

in which this generation was born was of great change and advance (see Fig. 1.3 for a selection of events between 1910 and 2010), particularly in such basic and important aspects like nutrition and sanitation, but also in significant socio historical, political and technological transformations. Over the past 100 years the world has changed, and its implications for human lives has been substantial. From two world wars, public health advances, the introduction of automobiles, radio and television in the first half of the twentieth century, to the contemporary and mainstream use of personal computers, internet and smartphones, centenarians have faced and face several challenges of adaptation.

Along with the physiological changes and social processes of ageing that are specific to individuals (age effects), also period effects and cohort effects play an important role in longevity. The first ones result from external factors that affect all age groups at a particular calendar time (e.g., war, famine, economic crisis), whereas the second ones correspond to variations that result from the unique experience/exposure of a cohort as they move across time. Either from an epidemiological or sociological point of view, both effects shape the experience of individual ageing as they are in the intersection of History and personal biographies. Ageing necessarily refers to a set of chronologically ordered events that simultaneously combine personal, group and historical markers: within these, life events can have positive/negative and proximal/distal effects in centenarians' health and overall quality of life (Martin & Martin, 2002).

Due to the societal improvements in the first half of the last century, a high proportion of early twentieth century cohorts survived to middle age. Indeed, birth

Table 1.1 Evolution and projections of number of centenarians by EU countries (1990–2050)

Country	1990	1995	2000	2005	2010	2015	2020	2025	2030	2035	2040	2045	2050
Austria	197	283	421	634	967	1,298	1,155	2,124	2,421	2,607	3,462	4,719	5,291
Belgium	409	570	826	1,087	1,419	1,769	1,885	3,304	4,112	4,909	5,479	6,045	8,598
Bulgaria	184	140	124	155	226	165	179	403	539	676	751	931	1,338
Croatia	74	95	146	211	403	387	236	378	564	718	910	970	1,296
Cyprus	16	16	21	30	49	77	99	131	193	260	371	508	719
Czechia	97	123	184	273	411	583	387	1,204	1,532	1,908	2,240	3,534	5,374
Denmark	343	434	553	698	902	1,140	1,326	1,708	1,937	2,188	2,932	4,133	6,111
Estonia	50	50	59	74	100	148	138	245	399	496	676	778	777
Finland	126	172	240	334	527	756	975	1,389	1,758	2,024	2,501	3,233	5,092
France	3,481	5,038	7,428	10,906	15,980	20,676	19,443	35,097	43,113	50,710	55,649	63,036	98,317
Germany	2,452	3,394	5,134	7,844	11,552	14,575	19,295	32,382	39,205	45,309	65,611	70,551	76,478
Greece	419	624	892	1,130	1,504	2,007	2,681	4,188	6,591	9,310	11,220	11,342	15,094
Hungary	154	203	292	388	567	861	618	1,747	1,893	2,477	3,022	3,907	4,878
Ireland	122	134	162	204	293	459	640	848	917	1,139	1,573	2,144	3,151
Italy	2,235	3,246	4,855	7,535	11,872	17,219	16,517	27,743	35,124	42,311	52,772	57,594	74,284
Latvia	138	145	140	164	189	386	559	1,239	2,002	2,375	2,381	3,141	2,721
Lithuania	498	413	267	234	259	247	571	1,832	3,572	4,935	5,460	6,191	5,789
Luxembourg	15	22	35	56	82	74	63	84	114	139	151	172	232
Malta	8	13	20	28	41	56	78	121	169	240	328	455	757
Netherlands	759	948	1,166	1,374	1,656	2,196	2,963	4,066	4,881	6,124	7,567	9,915	14,728
Poland	612	675	929	1,438	2,238	3,797	5,024	11,238	17,481	22,922	27,291	29,303	47,438

(continued)

Table 1.1 (continued)

Country	1990	1995	2000	2005	2010	2015	2020	2025	2030	2035	2040	2045	2050
Portugal	271	380	506	780	1,194	1,589	1,961	2,889	3,950	5,343	6,288	7,395	9,369
Romania	174	236	322	540	882	1,321	1,162	2,196	3,116	3,924	4,871	4,705	6,656
Slovakia	52	60	93	135	173	206	164	339	419	553	680	965	1,440
Slovenia	20	29	57	83	150	251	191	387	539	719	852	1,042	1,280
Spain	1,650	2,192	3,040	4,796	6,793	9,990	13,083	18,831	26,467	33,033	32,675	40,686	50,196
Sweden	555	706	914	1,182	1,522	1,945	2,270	2,863	3,025	3,264	3,954	5,613	7,526
United Kingdom	4,062	5,279	6,615	8,544	11,191	14,082	15,834	20,008	23,215	26,779	31,827	40,055	54,996
EU	19,173	25,620	35,441	50,857	73,142	98,260	109,497	178,984	229,248	277,392	333,494	383,063	509,926

Source UNdata | last update: 17 June 2019 | reference data: 1990–2050 | download date: 2 January 2020

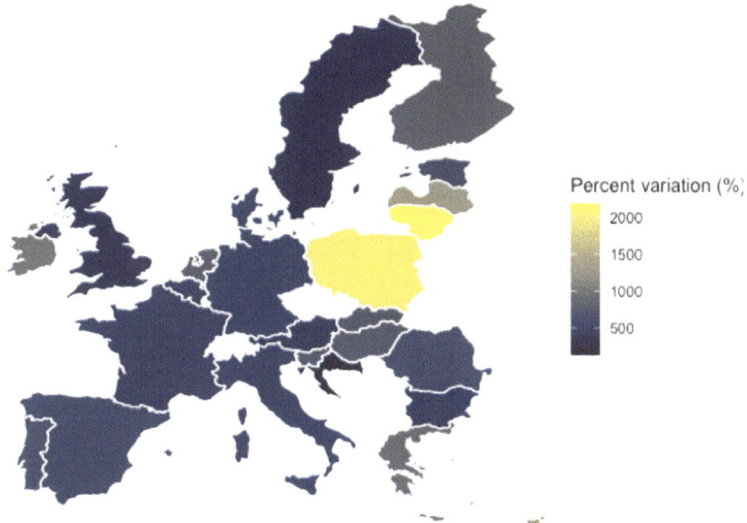

Fig. 1.2 Percent variation of number of centenarians in the period 2010–2050 in EU (*Source* UNdata | last update: 17 June 2019 | reference years: 2010 and 2050 | download date: 2 January 2020)

year cohort, as well as other factors like ethnicity, environment (industrialized/rural settings), education level and socioeconomic status, for instance, are important variables to take into account when describing the characteristics of those people identified as centenarians. Health progresses coming from effective disease prevention (e.g., vaccines, available drug therapies) and increased educational achievement are also particularly critical aspects to consider.

1.2 This Book's Objectives

This book focuses on the characteristics of those people identified as centenarians in several international available databases. The main purpose is to draw a profile of these individuals across EU countries. It starts by addressing some elementary aspects regarding the exceptional human longevity research (core concepts and approaches) and the methodological options we have assumed in our data analysis in what regards data sources and measures. Then, we continue by presenting the number of centenarians and supercentenarians by country and their distribution by sex. Educational levels attained by this population, their geographic distribution (rural vs. urban areas), marital status, elementary living situation (household type and household composition) and main causes of death are also described. After this description, an overview of main identified studies presently being conducted within the EU on centenarians is presented, namely their research focus and international collaborations. The book

Current centenarians were born	**1910s** In early 1910s the life expectancy in Europe was about 47 years old. World War I occurred throughout Europe (1914-1918). The Spanish Flue affected 500 million people and killed between 40-70 million (1918-1919). The first radio tuner system was patented (1917).
	1920s Widespread food pasteurization and refrigeration improved nutrition. Insulin was first administered to diabetics (1922), and first vaccines developed for diphtheria (1923), whooping cough (1926), tuberculosis (1927) and tetanus. A. Fleming discovered penicillin.
Centenarians were in their 20s	**1930s** First vaccine developed for yellow fever (1935) and typhus (1937). Civil War in Spain (1936-1939).
	1940s From 1939-1945 World War II occurred throughout the world. Penicillin became widely available to treat deadly diseases. First vaccine developed for influenza (1945). In 1946 United Nations was established.
Centenarians were in their 40s	**1950s** Polio was curbed as vaccines began to prevent life-threatening illnesses The European Common Market was established (1957).
	1960s Regulation of environmental pollution began, making air and water cleaner. Global period of economic growth – "The Swinging Sixties" in the Economic Union. In 1962 an artificial heart is used for the first time. In 1969 Apollo 11 lands on the moon.
Centenarians were in their 60s	**1970s** Effective retirement age (OECD average) in 1970 was 68.7 years for men and 66.5 years for women. First vaccine developed for pneumonia (1977). In 1977 British scientists report complete gene mapping of a living organism.
	1980s First vaccine developed for hepatitis B (1980). In 1981 IBM markets first personal computer and in 1983 compact discs (CD) are first sold. HIV, the virus that causes AIDS, is identified (1983). The first World Assembly on Ageing established the "Vienna International Plan of Action on Ageing" which called to specific action on health and nutrition, elderly consumers protections, housing and environment, education and research.
Centenarians in their 80's	**1990s** First vaccine developed for hepatitis A (1992). In 1990 East and West Germany are reunified. In 1991 USSR collapses and Cold Was ends. In 1992 Maastricht treaty created the European Union. Dolly the sheep becomes the first clone (1996).
	2000s Attack of the twin towers (2001) marked the coming of terrorism in the world arena. In 2002 the European Union released the EURO as a new form of currency. In 2009 the pandemic influenza A (H1N1) is estimated to have caused between 100,000 and 400,000 deaths globally in the first year alone. The second World Assembly on Ageing adopted a Political Declaration and the Madrid International Plan of Action on Ageing aimed at designing an international policy on ageing for the 21st century.
Reaching age 100	**2010s** European sovereign debt crisis marks first years of the 2010s, with several Eurozone member states particularly affected – Greece, Portugal, Ireland, Spain and Cyprus.

Fig. 1.3 Timeline of events—a selection (1910–2010)

ends with some final thoughts bearing in mind the need of further combining the eminently descriptive information presented with the information increasingly made available by each country's ongoing centenarian studies.

Chapter 2
Methodological Note

2.1 Measures of Exceptional Human Longevity

Research focused on extreme longevity is not new and the first papers concerning this topic appeared in the 40s, the majority in the field of medicine. Over the last decade, new ideas and opportunities regarding extreme longevity research emerged, namely through the lenses of diverse scientific approaches. One of the main issues in this research topic is the need to "quantify longevity", and within such need several measures have been proposed and used.

The first and simplest measure used to characterize and compare centenarian's populations is **prevalence**. This measure is the most commonly used in the literature given its facility to obtain by dividing the number of living centenarians in a given population by the total resident population. However, the results obtained with this formula can conduct to biased conclusions. For example, populations with immigrations flows in younger generations or a baby boom, resulting in an increase in the total resident population, will observe an underestimation of the prevalence of centenarians; in the other sense, populations that observed an emigration phenomenon, resulting in a decreasing number of the total resident population, will observe an overestimation of such prevalence. Therefore, migration and fertility indexes strongly influence this measure, and in these situations its use is not recommended (Herm, Cheung, & Poulain, 2012; Poulain, Herm, & Pes, 2013).

A second commonly used measure to study extreme longevity populations is **cohort life tables**. A life table is a table that shows, for each age, the probability of a person of that age to die before his/her next birthday. So, these tables use information about annual death statistics and age-specific population estimates on 1st January of each year. However, for oldest ages and specifically for centenarians, these tables cease to be reliable (Meslé & Vallin, 2002a), because the number of centenarians is small, which precludes firm estimation of the level and pattern of probability of dying (Modig, Andersson, Vaupel, Rau, & Ahlbom, 2017). Some authors have adapted the method to obtain life tables that are more accurate for the

© The Author(s), under exclusive license to Springer Nature Switzerland AG 2020
L. Teixeira et al., *Centenarians*, SpringerBriefs in Aging,
https://doi.org/10.1007/978-3-030-52090-8_2

oldest old (Meslé & Vallin, 2002a). Nevertheless, only a few number of countries produce cohort life tables including people 100+. Within Europe, examples include countries like France and Italy.

A third common indicator of longevity is **lifespan**. National institutes and researchers often use this measure, as it has the advantage of being quickly updated each year. It can be calculated for different ages (e.g., life expectancy at age 60), and the mostly used is the life expectancy at birth. Several publications revealed that life expectancy at birth has been increasing in the last decades. However, this increase could be related with the growing of the centenarian population and with the reduction in mortality below the age of 100 years, and not with the improvement in mortality amongst centenarians (Medford, Christensen, Skytthe, & Vaupel, 2019; Modig et al., 2017). As a matter of fact, although human life expectancy and median length of life rose dramatically over the past century, maximum life span has remained largely unchanged (Pignolo, 2019).

Over the last years, a set of measures has been increasingly used to explore and compare longevity populations. These include the **Extreme Longevity Index** (ELI) (Poulain et al., 2004), the **Centenarian Doubling Time** (CDT) (Robine & Caselli, 2005) and the **Centenarian Rate** (CR) (Robine & Caselli, 2005). The first one, ELI, is defined as the percentage of people born in a specific area (e.g., a country, a region or a city) who reached the age of 100 years old considering a specific cohort of interest. This index gives the probability that a person born in a given place has to become a centenarian; nevertheless, it presents the same pitfalls as the prevalence measure, given that it doesn't take into account fertility, infant mortality or migrations. CDT, on the other hand, represents the time trends in the annual rates of increase in the number of centenarians; it has emerged after some studies concluded that the number of centenarians doubles every ten years in low mortality countries since 1960s (Vaupel & Jeune, 1995). Finally, the CR was proposed to overcome the problems associated to the previous measures. This indicator is obtained by the ratio of the number of centenarians to the number of the people aged 60 years forty years before, expressed per 10,000. Considering as reference group (denominator) the people aged 60 years forty years earlier, the concerns with effects of fertility, infant mortality and migrations in a cohort are minimized, contributing to a more reliable longevity measure. Through the CR it is possible to compare countries and changes over time.

The measures previously described have been considered in several demographical studies, allowing obtaining valid and important information about population longevity and trends in the demographics of longevity. An important example is the case of the "Blue Zones" (BZ), a geographic clustering of exceptionally long-lived individuals. The concept, proposed in 2013 by Poulain et al. is defined as "a rather limited and homogeneous geographical area where the population shares the same lifestyle and environment and its longevity has been proved to be exceptionally high" (p. 89) (Poulain et al., 2013). Examples of validated blue zones are Ogliastra in Sardinia, Okinawa in Japan, the Nicoya peninsula in Costa Rica, and the island of Ikaria in Greece.

In terms of individual longevity, several examples can be reported and they often derive from well-known international records, as it is the case of the Guinness World Records. One the best well-known Guinness World Record on longevity was assigned to Jeanne Calment, a woman born in France in 1875 who has been considered as the longest-living person to date—she reached the age of 122 years and 164 days. Her age was exhaustively verified by several researchers (Jeune et al., 2010; Robine & Allard, 1999) and considered as the gold standard of the highest data quality for many decades in the scientific literature, despite several doubts have been expressed on her extreme outlying age (cf.: (Gavrilov & Gavrilova, 2000; Robine, Allard, Herrmann, & Jeune, 2019; Zak, 2019)). More illustrative cases of exceptional longevity include the case of the "first properly verified case of a 114-year-old man in human history" who was thought to be the "oldest man alive" back in 1996 (Wilmoth, Skytthee, Friou, & Jeune, 1996), and a recent record reached by two Portuguese brothers, Albano Andrade (born 14 December 1909) and Alberto Andrade (born 2 December 1911) who presented the highest combined age for two living siblings: their combined age as of 2 April 2019 was 216 years and 230 days (Ribeiro et al., 2020).

Regardless of the approach to be considered within demographical and individual longevity studies, an important note is to be given to **age validation,** as it is crucial in centenarians' studies. In fact, for measuring longevity the accurate identification of the age of the centenarian is a topic of incontrovertible importance (Herm et al., 2012), as inaccuracies in the measurement of extreme ages have been reported in populations without efficient civil registration. The more adequate attitude with this potential problem is to be positively critical, and develop adequate age validation exercises before launching larger research projects in order to identify populations with exceptional longevity (Poulain, 2010). The procedures of age validation are complex and comprise several steps that involve different sources such as consulting population registers, analyzing the identity card, and looking for reliable death records—if the person is dead. Given that data errors tend to increase with age, it is expected to have less accuracy of age information at 105+ years (Gavrilov & Gavrilova, 2019), with a need of extra attention in this group. More details about age validation can be obtained in Poulain (2010).

2.2 Database Sources, Concepts/Measures, and Statistical Analyses

In order to achieve this book's objectives, and considering the extent of measures that are available for reporting exceptional human longevity, some methodological options were taken and are addressed below. They mostly concern the list of countries from which data was retrieved and the reference year, data sources, core concepts/measures, scope search procedures and statistical analyses.

Countries and reference year

The reference year considered throughout this book was the year of 2018 (or nearest year). For some indicators, the reference year was last census (year 2011) as it was the only one presenting reliable information. Table 2.1 lists the EU countries here considered.[1]

Table 2.1 EU countries and acronyms

Country	Acronym
Austria	AT
Belgium	BE
Bulgaria	BG
Croatia	HR
Cyprus	CY
Czechia	CZ
Denmark	DK
Estonia	EE
Finland	FI
France	FR
Germany	DE
Greece	EL
Hungary	HU
Ireland	IE
Italy	IT
Latvia	LV
Lithuania	LT
Luxembourg	LU
Malta	MT
Netherlands	NL
Poland	PL
Portugal	PT
Romania	RO
Slovakia	SK
Slovenia	SI
Spain	ES
Sweden	SE
United Kingdom	UK

---

[1] As of 1st February 2020, the United Kingdom is no longer part of the EU. However, given that the book use information from previous years, the authors opted to maintain UK as EU member.

Sources

This book relies on official data sources, exception comprising the International Database on Longevity, the Gerontology Research Group, and the Human Mortality Database. The option to use this kind of sources (and no other studies) is related with the representativeness of centenarians (normally, this specific group is underrepresented) and/or the absence of studies for some EU countries.

In overall, data used in each chapter came from six different sources:

- **CENSUS HUB**: the 2011 Census database is the result of a major joint effort by the European Statistical System (ESS) to better disseminate the results of the Population and Housing Censuses in Europe; it provides users with easy access to detailed census data that are structured in the same way, and can be methodologically comparable between countries (European Comission, 2014).

  https://ec.europa.eu/CensusHub2/query.do?step=selectHyperCube&qhc=false.

- **International Database on Longevity (IDL)**: The IDL is the result of an ongoing, concerted effort to provide thoroughly validated information on individuals who have attained extreme ages. The IDL allows for the demographic analysis of mortality at the highest ages, and currently has information on supercentenarians for 13 countries: 11 European countries, as well as for Canada (Quebec only), and the United States of America (International Database on Longevity, 2020).

  https://www.supercentenarians.org.

- **Gerontology Research Group (GRG)**: the GRG is an important authority on the world's oldest humans, specifically supercentenarians research, authentication and database (Gerontology Research Group, 2020).

  http://www.grg.org/SC/WorldSCRankingsList.html.

- **Human Mortality Database (HMD)**: the HMD was created to provide detailed mortality and population data. This database contains original calculations of death rates and life tables for national populations (countries or areas), as well as the input data used in constructing those tables. The input data consist of death counts from vital statistics, plus census counts, birth counts, and population estimates from various sources (Human Mortality Database, 2020).

  https://www.mortality.org/.

- **UNdata**: UNdata is a web-based data service for the global user community that assemble a variety of statistical resources compiled by the United Nations (UN) statistical system and other international agencies (United Nations, 2019b).

  http://data.un.org/.

- **World Health Organization (WHO) Mortality Database**: the WHO Mortality Database is a compilation of mortality data by age, sex and cause of death, as reported annually by Member States from their civil registration systems (World Health Organization, 2020).

https://www.who.int/healthinfo/mortality_data/en/.

Core concepts and measures

- **Centenarian**: Person with 100+ years old

- **Supercentenarian**: Person with 110+ years old

- **Centenarian Ratio (CR)**: Ratio of the number of people aged 100 years to the number of the people aged 60 years, forty years before, expressed per 10,000 (Robine & Caselli, 2005). In this book, considering 2018 as the reference year, CR is the ratio of the number of people with 100 in 2018 to the number of people with 60 years in 1978, expressed by 10,000. For some countries, information was not available for 2018, and the nearest year was considered: 2017 for Denmark, Netherlands, Poland, Spain and UK; 2016 for Portugal; and 2015 for Italy. Seven countries were excluded because some information was not available: Croatia, Cyprus, Germany, Greece, Malta, Romania and Slovenia. CR was calculated for the total population and by sex:

$$CR = \left(\frac{Number\ of\ people\ aged\ 100\ years\ in\ 2018/}{Number\ of\ people\ aged\ 60\ years\ in\ 1978} \right) \times 10,000$$

$$CR_{male} = \left(\frac{Number\ of\ male\ aged\ 100\ years\ in\ 2018/}{Number\ of\ male\ aged\ 60\ years\ in\ 1978} \right) \times 10,000$$

$$CR_{female} = \left(\frac{Number\ of\ female\ aged\ 100\ years\ in\ 2018/}{Number\ of\ female\ aged\ 60\ years\ in\ 1978} \right) \times 10,000$$

- **Sex Ratio (RR_{sex})**: Rate between CR_{female} and CR_{male}. In this work, the 99% confidence interval for the RR_{sex} was also calculated.

$$RR_{sex} = CR_{female}/CR_{male}$$

- **International Standard Classification of Education (ISCED)**: Reference international classification for organizing education programs and related qualifications by levels and fields. For ISCED 1997, 7 levels were defined: ISCED Level 0. No formal education; ISCED Level 1. Primary education; ISCED Level 2. Lower secondary education; ISCED Level 3. Upper secondary education; ISCED Level 4. Post-secondary non-tertiary education; ISCED Level 5. First stage of tertiary education; ISCED Level 6. Second Stage of tertiary education (Eurostat, 2018). In this book, levels 3 into 6 were aggregated.

- **Urban/Rural**: Because of national differences in the characteristics that distinguish urban from rural areas, the distinction between the urban and the rural population is not yet amenable to a single definition that would be applicable to all countries or, for the most part, even to the countries within a region (United Nations, 2017). The definitions used by individual countries can be conferred at the 2018 Demographic Yearbook from the United Nations (United Nations, 2019a). In this work, the category "urban" was the one selected to be explored and the 99% confidence interval for the proportion was calculated.

- **Marital status**: civil (or marital) status of an individual, i.e., the legal, conjugal status of an individual in relation to the marriage laws or customs of the country (single, married, widowed and not remarried, divorced and not remarried, married but separated) (United Nations, 2017). In this work, the categories "divorced and not remarried" and "married but separated" were merged in one category "divorced and not remarried or married but separated", and the categories "single (never married)" and "married" were further explored by sex.

- **Household**: A household may be located in a housing unit or in a set of collective living quarters such as a boarding house, a hotel or a camp, or may comprise the administrative personnel in an institution. The household may also be homeless (United Nations, 2017). According to the UN (United Nations, 2017), households should be classified by type according to the number of family nuclei they contain and the relationship, if any, between the family nuclei and the other members of the household. The relationship should be through blood, adoption or marriage, to whatever degree is considered pertinent by the country. The types of household to be distinguished could be: (a) One-person household; (b) Nuclear household; (c) Extended household, (d) Composite household; (e) Other/Unknown. In this book, the categories "collective living quarters" and "one-person household" were further explored by sex.

- **International Classification of Diseases (ICD)**: ICD-10 defines the universe of diseases, disorders, injuries and other related health conditions. This classification was used for the cause of death information (World Health Organization, 2011). In this book, information for the main causes of death was described for people with 95+ by sex (information for 100+ was not available).

Scope search procedures
The databases Web of Science[2] and Scopus[3] were searched (8th May 2020) seeking to map existing literature on centenarian studies in terms of its nature, features and volume over the last twenty years (2000–2020). The key term considered in the search was "centenarian" (title, abstract, keyword) and each one of the previously listed

---

[2]Web of Science is a publisher-independent global citation database (https://clarivate.com/webofsciencegroup/solutions/web-of-science/).

[3]Scopus is a source-neutral abstract and citation database curated by independent subject matter experts (https://www.elsevier.com/solutions/scopus).

countries (affiliation or address). Only publications in English language and referring to human longevity were selected, i.e., documents reporting topics on Archaeology, Architecture or Agriculture were excluded. A summary of findings is presented as figures and/or tables, based on three main topics:

- **Year of publication**

- **Country**
  Information about the countries involved in a publication was obtained based on the address of the authors (Web of Science) or on the affiliation of the authors (Scopus).

- **Research area**
  Research area classifications are different in the two sources considered. In the Web of Science database, research areas are classified into five broad categories: Arts & Humanities, Life Sciences & Biomedicine (which includes Geriatrics and Gerontology), Physical Sciences, Social Sciences, and Technology. In Scopus, there are four main subject areas: Health Sciences, Life Sciences (which includes a second level subject field entitled "Biochemistry, Genetics and Molecular Biology" that includes the topic "Aging"), Physical Sciences, and Social Sciences.

Statistical analyses

Throughout the book, descriptive analyses of the measures selected for each chapter and described above are presented using tables and/or figures, by EU country. Additionally, 99% confidence intervals for RR_{sex} (Sect. 3.2) and for proportion of centenarians that lived in urban areas (Sect. 3.4) were calculated. Chi-Square test or Fisher exact test was performed in order to evaluate the association between sex and household (considering 0.01 as level of significance), by EU country (Sect. 3.6).

For some topics and/or countries, information was not available or was not considered reliable (information about data reliability was provided by each source). In such cases, the country was excluded if information "not available" and/or "not reliable" was higher than 5% of the total information. For each topic, and considering the countries with available data, information about the total centenarian population (TOT) was also considered. Table 2.2 lists the countries included in each chapter, according to the information available.

Table 2.3 presents the sources used in each chapter.

Table 2.2 EU countries included in each chapter

Country	3.1 Cents and supercents	3.2 Sex	3.3 Education	3.4 Rural/Urban	3.5 Marital status	3.6 Household	3.6 Type of household	3.7 Causes of death Main cause	3.7 Causes of death %
Austria	•	•	•	•	•	•	•	•	•
Belgium	•	•		•	•	•		•	•
Bulgaria	•	•		•	•	•			
Croatia			•	•		•	•	•	
Cyprus			•						
Czechia	•	•		•	•	•	•	•	•
Denmark	•	•		•	•	•		•	•
Estonia	•	•		•	•	•	•	•	•
Finland	•	•	•					•	•
France	•		•					•	
Germany				•	•	•	•	•	•
Greece			•	•	•	•	•	•	
Hungary	•	•		•	•	•	•	•	
Ireland	•	•		•	•	•	•	•	
Italy	•	•	•			•	•	•	
Latvia	•	•	•	•	•	•	•	•	•
Lithuania	•	•			•	•	•	•	•
Luxembourg	•	•						•	•

(continued)

Table 2.2 (continued)

Country	Chapter: 3.1 Cents and supercents	3.2 Sex	3.3 Education	3.4 Rural/Urban	3.5 Marital status	3.6 Household: Household	Type of household	3.7 Causes of death: Main cause	%
Malta			•	•	•	•	•	•	•
Netherlands	•	•						•	•
Poland	•	•		•		•	•	•	
Portugal	•	•	•	•	•	•	•		
Romania	•		•	•				•	
Slovakia	•	•	•	•	•				
Slovenia			•	•	•				
Spain	•	•			•	•	•	•	•
Sweden	•	•						•	•
United Kingdom	•	•	•	•	•	•	•	•	

Table 2.3 Sources used in each chapter

Source	Chapter						
	3.1 Cents and supercents	3.2 Sex	3.3 Education	3.4 Rural/Urban	3.5 Marital status	3.6 Household	3.7 Causes of death
CENSUS HUB			•				
IDL	•						
GRG	•						
HMD	•	•					
UNdata				•	•	•	
WHO mortality database							•

Chapter 3
Profiling European Centenarians

3.1 The Increasing Number of (Super) Centenarians

The number of centenarians has been rising all over EU in the last decades. This phenomenon of longevity can be analyzed comparing different countries, as possible through the CR presented in Fig. 3.1 and Table 3.1.

The total CR value is 107, which corresponds to 30,053 person aged 100 years to 2,815,868 persons aged 60 of the same cohort. Only three countries have ratios of reaching age 100 years above the total and higher than 100: France with a CR equal to 188, followed by Spain and Italy, with a CR equal to 147 and 128, respectively.

The high proportion of centenarians observed in some countries is largely the result of their lower mortality rates and older population age structures. Along with having the largest populations, in these three countries life expectancy is particularly high: 82.7 for France, 83.4 for Spain, and 83.1 for Italy, whereas for the EU is 80.9 years old (Eurostat, 2019). Another common aspect to the three countries is the geographical location in the south and Mediterranean zone of Europe, which has been considered a longevity zone due to the Mediterranean lifestyle, especially in terms of diet (Martinez-Gonzalez & Martín-Calvo, 2016), as well as related with the French paradox, i.e., the lower-than-expected coronary heart disease mortality (De Lorgeril et al., 2002). Although this analysis is focused on country differences, several authors have been drawing attention to the high variability within the same country across different zones and regions, which is particularly frequent on these southern countries (Caselli & Lipsi, 2016; Gómez-Redondo & González, 2010).

At the other extreme in terms of CR are Bulgaria, Slovakia and Czechia with a CR between 15 and 31. The lower ratios of centenarians in countries of the former east-bloc and, in some cases, post-Soviet states can be associated to a general crisis that considerably hampered their progress, namely in health politics, and life expectancies that dropped steeply in the 1990s (Meslé & Vallin, 2002b). Cardiovascular diseases are more frequent in these countries due to higher consumption of alcohol

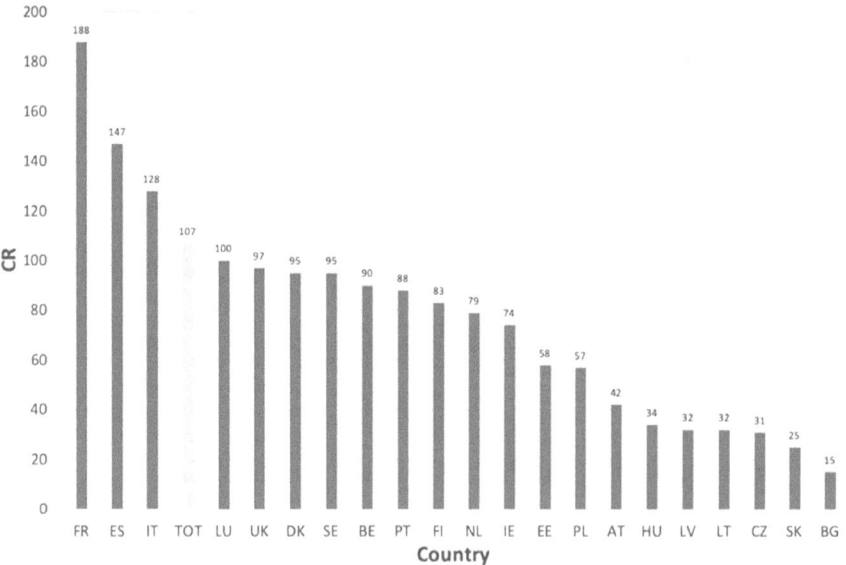

Fig. 3.1 Centenarian ratio by EU country (*Source* Human Mortality Database and External Sources (Estimates are original creations of the Human Mortality Database (Human Mortality Database, 2020), with exception of estimates of people aged 60 years for some countries, where estimates came directly from external sources, namely: Austria (Statistik Austria, 2020), Bulgaria (National Statistical Institute, 2020), Denmark (Danmarks Statistik, 2020), Finland (Statistics Finland, 2020), France (Vallin & Meslé, 2001), Latvia (Central Statistical Bureau (CSB) of Latvia, 2020), Netherlands (Centraal Bureau voor de Statistiek, 2020), Slovakia (Statistical Office of the Slovak Republic, 2020) and Sweden (Lundström, 2020).) | last update: according to the country (Austria: 03/09/2018; Belgium: 06/09/2019; Bulgaria: 27/09/2019; Croatia: 14/09/2018; Czechia: 25/03/2019; Denmark: 23/04/2018; Estonia: 20/09/2018; Finland: 02/12/2019; France: 01/11/2019; Germany: 17/12/2018; Greece: 17/04/2018; Hungary: 26/09/2018; Ireland: 01/10/2019; Italy: 26/09/2017; Latvia: 20/09/2018; Lithuania: 20/09/2018; Luxembourg: 10/12/2019; Netherlands: 10/05/2018; Poland: 30/08/2018; Portugal: 26/09/2017; Slovakia: 06/12/2018; Slovenia: 06/12/2018; Spain: 05/10/2018; Sweden: 06/02/2019; United Kingdom: 28/05/2018.) | reference year: 2018 (or nearest year) | download date: 2 January 2020)

and tobacco. The breakdown of social safety nets, rising social inequality, and inadequate health services are pointed as important causes of a lower life expectancy (Rechel et al., 2013).

Researches from different disciplines (e.g., demography, epidemiology, biomedical sciences) have been proving that the higher rates of centenarians are mainly related to the decrease in mortality at old age (Robine & Caselli, 2005). On this matter, Vaupel (2010) adds that death is being delayed because people are reaching old age in better health. The author also draws attention to the importance of remarkable progresses, as the declines in mortality at early ages observed at the beginning of the twentieth century, the reduction in mortality from heart disease during the last third of that century, and the declining mortality in old age in recent decades (Vaupel, 2010). It is important to refer, however, that recent studies on the pattern of mortality above the age of 100 years suggest that it is not changing despite improvements at younger ages (Modig et al., 2017).

Along with the increasing number of individuals reaching the boundary of 100 years old, great interest has arisen around those who have reached an ever more exceptional age—supercentenarians, i.e., individuals who had reached 110 years of age. For this particular population, more than the inquisitiveness on factors

Table 3.1 Centenarian Ratio (CR) by EU country

Country	No. of persons with an exact age of 60 years in 1978[a]	No. of persons with an exact age of 100 years in 2018[a]	Centenarian Ratio (CR)
Austria	92,884	394	42
Belgium	64,817	581	90
Bulgaria	42,533	65	15
Croatia	NA	62	–
Cyprus	NA	NA	–
Czechia	56,546	176	31
Denmark	53,045	505	95
Estonia	8,861	51	58
Finland	44,117	365	83
France	300,721	5,641	188
Germany	NA	3,651	–
Greece	NA	NA	–
Hungary	63,484	215	34
Ireland	26,778	197	74
Italy	603,006	7,698	128
Latvia	15,527	49	32
Lithuania	18,455	59	32
Luxembourg	2,889	29	100
Malta	NA	NA	–
Netherlands	121,432	959	79
Poland	206,064	1,178	57
Portugal	89,334	782	88
Romania	NA	NA	–
Slovakia	21,498	54	25
Slovenia	NA	52	–
Spain	306,992	4,523	147
Sweden	95,498	907	95
United Kingdom	581,387	5,625	97

(continued)

Table 3.1 (continued)

Country	No. of persons with an exact age of 60 years in 1978[a]	No. of persons with an exact age of 100 years in 2018[a]	Centenarian Ratio (CR)
Total	2,815,868	30,053	107

NA Data not available
Source Human Mortality Database and External Sources[b] | last update: according to the country[c] | reference year: 2018 (or nearest year) | download date: 2 January 2020
[a]See methodological note
[b]Estimates are original creations of the Human Mortality Database (Human Mortality Database, 2020), with exception of estimates of people aged 60 years for some countries, where estimates came directly from external sources, namely: Austria (Statistik Austria, 2020), Bulgaria (National Statistical Institute, 2020), Denmark (Danmarks Statistik, 2020), Finland (Statistics Finland, 2020), France (Vallin & Meslé, 2001), Latvia (Central Statistical Bureau (CSB) of Latvia, 2020), Netherlands (Centraal Bureau voor de Statistiek, 2020), Slovakia (Statistical Office of the Slovak Republic, 2020) and Sweden (Lundström, 2020)
[c]Austria: 03/09/2018; Belgium: 06/09/2019; Bulgaria: 27/09/2019; Croatia: 14/09/2018; Czechia: 25/03/2019; Denmark: 23/04/2018; Estonia: 20/09/2018; Finland: 02/12/2019; France: 01/11/2019; Germany: 17/12/2018; Greece: 17/04/2018; Hungary: 26/09/2018; Ireland: 01/10/2019; Italy: 26/09/2017; Latvia: 20/09/2018; Lithuania: 20/09/2018; Luxembourg: 10/12/2019; Netherlands: 10/05/2018; Poland: 30/08/2018; Portugal: 26/09/2017; Slovakia: 06/12/2018; Slovenia: 06/12/2018; Spain: 05/10/2018; Sweden: 06/02/2019; United Kingdom: 28/05/2018

associated with their very long lives (Maier, Gampe, Jeune, Robine, & Vaupel, 2010), particular attention has been given to the way they live the period between 100 years of age and death. In fact, this has become a key issue for researchers interested in the study of these rare and special cases.

Different sources can be used to obtain information about supercentenarians, namely the IDL, the GRG and the HMD. The first two sources give information at individual level (considering age validation procedures to identify supercentenarians), whereas the third one provides information at population level (considering information about population and/or death for specific cohorts).

The IDL, which is an international collaborative group of researchers dedicated to gather validated demographic data on supercentenarians, currently presents the greatest amount of data regarding these extreme long-lived individuals. The first cases of validated supercentenarians appeared in the 1960s but their numbers have steadily increased since the mid 1980s. The IDL currently has information on supercentenarians for 13 countries, of which 9 are within the EU. The candidates for inclusion in the IDL database are from records of government agencies based on records of deaths obtained from the vital registration system. These candidates meet strict criteria for the validity of their age and, for this reason, the IDL does not provide exhaustive sets of validated supercentenarians for any country.

Figure 3.2 presents information obtained from IDL about supercentenarians identified in the period 2011–2017 by sex. According to this source, 216 supercentenarians lived in the EU between 2011 and 2017, 125 from France (3 male and 122 female), 65 from the UK (4 male and 61 female), and the other from Austria

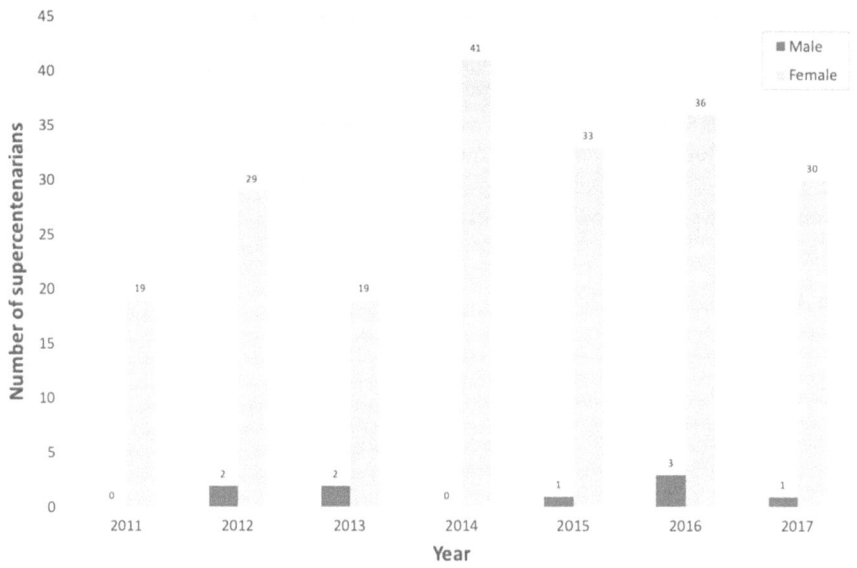

Fig. 3.2 Number of supercentenarians by year (*Source* International Database on Longevity | last update: 09 December 2019 | reference year: 2011–2017 | download date: 13 January 2020)

(4 female), Belgium (1 male and 6 female), Denmark (1 female), and Spain (1 male and 13 female).

The GRG, credited as a contributor to the IDL, also has information regarding validated supercentenarians. The age validation of the supercentenarians considers multiple documents throughout their life course to prove their age, including, at minimum, three documents: one early-life document, one mid-life document, and one late-life document. The list of supercentenarians provided by this source does not include all living supercentenarians, only those whose claim can be validated to the required standard. On February 2019, according to this source, the number of living supercentenarians was 35 (all female), of which 12 were from EU countries (6 from Italy, 2 from France, 2 from Netherlands, 1 from UK and 1 from Poland) (Young & Kroczek, 2019).

Additionally, an estimation of the number of supercentenarians could be obtained using the population estimates, based on the HMD (Human Mortality Database, 2020); such information is presented in Table 3.2. Based on these estimates, a total of 112 supercentenarians could be considered in 2018 (or nearest), distributed by gender and EU country in the following way: 50 in France (1 male and 49 female); 15 female in UK; 13 female in Italy; 13 in Spain (2 male and 11 female); 10 female in Germany; 5 in Poland (1 male and 4 female); 2 female in Sweden; and 1 female in Denmark, Greece, Hungary and Portugal.

In low-mortality countries, estimate of probability of death between age 110 through 114 is stable and around 50% (Robine & Vaupel, 2002). Although super-centenarians aged slowly, they became extremely frail in their final years and their physical functions declined markedly, especially after their 105th birthdays (Maier et al., 2010). They tend to present high prevalence of conditions such as cataract, osteoporosis and bone fractures but low prevalence of more significant chronic degenerative pathologies, coronary heart disease, stroke and cancer (Vacante et al., 2012). Therefore, although they exhibited multiple clinical markers of frailty coincident with a rapid Activity of Daily Living (ADL) decline in their last years of life, they present an elite phenotype (Willcox et al., 2008).

Table 3.2 Supercentenarians by EU country

Country	Male	Female	Total
Austria	0	0	0
Belgium	0	0	0
Bulgaria	0	0	0
Croatia	0	0	0
Cyprus	NA	NA	NA
Czechia	0	0	0
Denmark	0	1	1
Estonia	0	0	0
Finland	0	0	0
France	1	49	50
Germany	0	10	10
Greece	0	1	1
Hungary	0	1	1
Ireland	0	0	0
Italy	0	13	13
Latvia	0	0	0
Lithuania	0	0	0
Luxembourg	0	0	0
Malta	NA	NA	NA
Netherlands	0	0	0
Poland	1	4	5
Portugal	0	1	1
Romania	NA	NA	NA
Slovakia	0	0	0
Slovenia	0	0	0
Spain	2	11	13
Sweden	0	2	2
United Kingdom	0	15	15

(continued)

Table 3.2 (continued)

Country	Male	Female	Total
Total	4	108	112

NA Data not available

Source Human Mortality Database and External Sources[a] | last update: according to the country[b] | reference year: 2018 (or nearest year) | download date: 14 January 2020

[a]Estimates are original creations of the Human Mortality Database (Human Mortality Database, 2020), with exception of estimates of people aged 60 years for some countries, where estimates came directly from external sources, namely: Austria (Statistik Austria, 2020), Bulgaria (National Statistical Institute, 2020), Denmark (Danmarks Statistik, 2020), Finland (Statistics Finland, 2020), France (Vallin & Meslé, 2001), Latvia (Central Statistical Bureau (CSB) of Latvia, 2020), Netherlands (Centraal Bureau voor de Statistiek, 2020), Slovakia (Statistical Office of the Slovak Republic, 2020) and Sweden (Lundström, 2020)

[b]Austria: 03/09/2018; Belgium: 06/09/2019; Bulgaria: 27/09/2019; Croatia: 14/09/2018; Czechia: 25/03/2019; Denmark: 23/04/2018; Estonia: 20/09/2018; Finland: 02/12/2019; France: 01/11/2019; Germany: 17/12/2018; Greece: 17/04/2018; Hungary: 26/09/2018; Ireland: 01/10/2019; Italy: 26/09/2017; Latvia: 20/09/2018; Lithuania: 20/09/2018; Luxembourg: 10/12/2019; Netherlands: 10/05/2018; Poland: 30/08/2018; Portugal: 26/09/2017; Slovakia: 06/12/2018; Slovenia: 06/12/2018; Spain: 05/10/2018; Sweden: 06/02/2019; United Kingdom: 28/05/2018

3.2 The Feminization of Exceptional Longevity

Universally, women live longer than man (Pignolo, 2019) and female centenarians largely outnumber male centenarians. Within the EU, and in line with the total population, France, Italy and Spain have the highest ratios for both sexes (Table 3.3). Nevertheless, the CR of male is the highest in Spain (CR = 63) and the CR of female is the highest in France (CR = 301).

Figure 3.3 presents information about the rate of centenarian ratio (female/male) by EU country. As expected, the CR is always significantly higher in women comparing with men ($RR_{sex} > 1$ for all countries). The total RR_{sex} is of 4.23, meaning that there were 4.23 women for each man that reached the age of 100 years. Excluding Luxembourg and Estonia (where RR_{sex} are overestimated given the small number of men), the country that presents the highest ratio between sexes concerning CR is Denmark (5.17), followed by Netherlands (5.14) and France (5.03). All of these countries have at least 5 women for each man, i.e., 80% female to 20% male. In countries like Bulgaria and Lithuania the ratio between CR_{female} and CR_{male} is lower than 3 (i.e., more than 66% of people reaching 100 are female).

Women have higher odds of achieving 100 years old and this tendency follows the difference in life expectancy, a European and even worldwide phenomenon indicating that human longevity seems strongly influenced by sex (Barford & Dorling, 2006).

Table 3.3 Centenarians Ratio (CR) by sex and rate of CR (RR_{sex}) and 99% confidence interval (CI) by EU country

Country	No. of female with an exact age of 60 years in 1978[a]	No. of female with an exact age of 100 years in 2018[a]	CR_{female}	No. of male with an exact age of 60 years in 1978[a]	No. of male with an exact age of 100 years in 2018[a]	CR_{male}	Rate of CR (RR_{sex})
Austria	52,779	323	61	40,105	72	18	3.40
Belgium	34,342	488	142	30,475	94	31	4.61
Bulgaria	22,028	44	20	20,505	20	10	2.05
Croatia	NA	47	–	NA	15	–	–
Cyprus	NA	NA	–	NA	NA	–	–
Czechia	30,310	148	49	26,237	28	11	4.58
Denmark	27,283	427	157	25,762	78	30	5.17
Estonia	5,467	48	88	3,393	3	9	9.93
Finland	25,478	311	122	18,639	54	29	4.21
France	159,376	4,796	301	141,345	845	60	5.03
Germany	NA	3,128	–	NA	524	–	–
Greece	NA	NA	–	NA	NA	–	–
Hungary	34,453	172	50	29,031	43	15	3.37
Ireland	13,433	162	121	13,345	35	26	4.60
Italy	320,916	6,349	198	282,090	1,349	48	4.14
Latvia	9,426	42	45	6,101	6	10	4.53
Lithuania	10,937	47	43	7,519	12	16	2.69
Luxembourg	1,523	27	177	1,366	2	15	12.11
Malta	NA	NA	–	NA	NA	–	–
Netherlands	63,872	816	128	57,560	143	25	5.14
Poland	115,547	958	83	90,517	219	24	3.43
Portugal	47,599	639	134	41,734	143	34	3.92
Romania	NA	NA	–	NA	NA	–	–
Slovakia	11,387	43	38	10,111	11	11	3.47
Slovenia	NA	44	–	NA	8	–	–
Spain	168,066	3,647	217	138,926	876	63	3.44
Sweden	48,683	745	153	46,815	162	35	4.42
United Kingdom	306,059	4,614	151	275,328	1,011	37	4.11

(continued)

Table 3.3 (continued)

Country	No. of female with an exact age of 60 years in 1978[a]	No. of female with an exact age of 100 years in 2018[a]	CR_{female}	No. of male with an exact age of 60 years in 1978[a]	No. of male with an exact age of 100 years in 2018[a]	CR_{male}	Rate of CR (RR_{sex})
Total	1,474,683	24,846	168	1,306,904	5,206	40	4.23

NA Data not available
Source Human Mortality Database and External Sources[b] | last update: according to the country[c] | reference year: 2018 (or nearest year) | download date: 2 January 2020
[a]See methodological note
[b]Estimates are original creations of the Human Mortality Database (Human Mortality Database, 2020), with exception of estimates of people aged 60 years for some countries, where estimates came directly from external sources, namely: Austria (Statistik Austria, 2020), Bulgaria (National Statistical Institute, 2020), Denmark (Danmarks Statistik, 2020), Finland (Statistics Finland, 2020), France (Vallin & Meslé, 2001), Latvia (Central Statistical Bureau (CSB) of Latvia, 2020), Netherlands (Centraal Bureau voor de Statistiek, 2020), Slovakia (Statistical Office of the Slovak Republic, 2020) and Sweden (Lundström, 2020)
[c]Austria: 03/09/2018; Belgium: 06/09/2019; Bulgaria: 27/09/2019; Croatia: 14/09/2018; Czechia: 25/03/2019; Denmark: 23/04/2018; Estonia: 20/09/2018; Finland: 02/12/2019; France: 01/11/2019; Germany: 17/12/2018; Greece: 17/04/2018; Hungary: 26/09/2018; Ireland: 01/10/2019; Italy: 26/09/2017; Latvia: 20/09/2018; Lithuania: 20/09/2018; Luxembourg: 10/12/2019; Netherlands: 10/05/2018; Poland: 30/08/2018; Portugal: 26/09/2017; Slovakia: 06/12/2018; Slovenia: 06/12/2018; Spain: 05/10/2018; Sweden: 06/02/2019; United Kingdom: 28/05/2018

One main issue when discussing sex differences in longevity is the role played by estrogen and progesterone. Menopause reflects the inevitable final hallmark of a woman's fertile lifespan and of the beneficial effects of estrogens on immune responses. Rapid reduction of estrogen levels results in an increased susceptibility and mortality toward a series of infectious diseases caused in old women losing of their immunological privilege toward infection. It is noteworthy that women will soon spend half of their life in post-menopause, if the current trend of increasing human life expectancy persists (Bozzini & Falcone, 2017). The particular protective role of estrogen that ends its role with menopause probably explains the raised risk of (cardiovascular) disease in older women. However, what we know and observe looking at survival curves and healthy survival is not sufficient to provide a full explanation of the survival gap and its evolution. In fact, the life expectancy until the second half of the XX century did not present such a gap and men and women had very similar life expectancies. In 2010, in EU countries, life expectancy at age 50 was 29.8 years for men and 34.6 years for women, but the average duration of healthy life was similar between sexes (68.6 years for women, and 67.9 years for men). This means that most of the 5 years of women advantage in life expectancy are years with disease and disability.

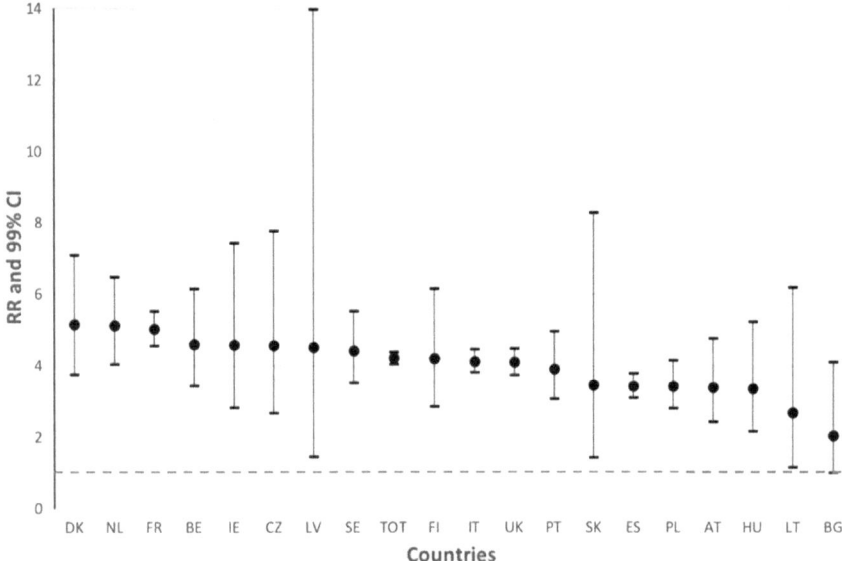

Fig. 3.3 Rate of centenarian ratio (RR_{sex}) and 99% confidence interval (CI) by EU country (*Source* Human Mortality Database and External Sources (Estimates are original creations of the Human Mortality Database (Human Mortality Database, 2020), with exception of estimates of people aged 60 years for some countries, where estimates came directly from external sources, namely: Austria (Statistik Austria, 2020), Bulgaria (National Statistical Institute, 2020), Denmark (Danmarks Statistik, 2020), Finland (Statistics Finland, 2020), France (Vallin & Meslé, 2001), Latvia (Central Statistical Bureau (CSB) of Latvia, 2020), Netherlands (Centraal Bureau voor de Statistiek, 2020), Slovakia (Statistical Office of the Slovak Republic, 2020) and Sweden (Lundström, 2020).) I last update: according to the country (Austria: 03/09/2018; Belgium: 06/09/2019; Bulgaria: 27/09/2019; Croatia: 14/09/2018; Czechia: 25/03/2019; Denmark: 23/04/2018; Estonia: 20/09/2018; Finland: 02/12/2019; France: 01/11/2019; Germany: 17/12/2018; Greece: 17/04/2018; Hungary: 26/09/2018; Ireland: 01/10/2019; Italy: 26/09/2017; Latvia: 20/09/2018; Lithuania: 20/09/2018; Luxembourg: 10/12/2019; Netherlands: 10/05/2018; Poland: 30/08/2018; Portugal: 26/09/2017; Slovakia: 06/12/2018; Slovenia: 06/12/2018; Spain: 05/10/2018; Sweden: 06/02/2019; United Kingdom: 28/05/2018.) I reference year: 2018 (or nearest year) I download date: 2 January 2020)

Franceschi et al. (2000) compared trajectories of extreme longevity in Italy according to sex, and concluded that centenarians are a very heterogenous group, with marked sex differences: male centenarians tend to be less heterogenous and healthier than female centenarians (Franceschi et al., 2000). There is evidence of genetic factors (e.g., mtDNA haplogroups, Thyrosine Hydroxilase, and IL-6 genes) being responsible for extreme longevity, but it is also recognized the role of different strategies used by men and women throughout them to achieve longevity. It seems that female longevity is less dependent on genetics than male longevity, and that female centenarians' profit from healthier life-style and favorable environment. Anyway, the major gap between sexes may sat principally in the length of life than in healthy longevity as according to Ostan et al. (2016).

Inequality in exceptional longevity is also associated with wage and Neumayer & Plumper (2016) reported that income inequality before transfers was positively associated with inequality in longevity; income redistribution, on the other hand, was negatively associated with longevity inequality (Neumayer & Plumper, 2016). This conclusion points to the idea that reducing longevity inequality may be achieved through public health policies, and also via the reduction of income inequality and the redistribution of incomes. In line with this view, Barthold et al. (2014) reported that health-spending is associated with greater life expectancy, more accentuated for men than women, in almost all OECD (Organization for Economic Co-operation and Development) countries, especially when life expectancy is calculated at 65 years (Barthold, Nandi, Mendoza Rodríguez, & Heymann, 2014).

3.3   The Diversity of Educational Attainment

An important component of social factors and status is education attainment, which is known to influence a set of individual characteristics, have a great dependence of contextual factors, and to be associated with many outcomes, including longevity. Particularly for the centenarian population, studies tend to report some diversity in terms of their profile in what regards their most common educational levels, which can also be observed throughout the EU countries here considered. As it can be seen in Fig. 3.4 and Table 3.4, Portugal, Spain and Greece are the countries with higher percentage of centenarians without formal education, with values varying between 61.60% and 52.69%. Even the ones who had attended school did not reach high schooling levels, since only 59 centenarians (3.87%) in Portugal, 306 (12.30%) in Greece, and 570 (6.96%) in Spain have attained upper secondary education or a higher level. On the opposite side, Finland and the United Kingdom are the countries with higher educational levels, with all centenarians presenting at least a level 2 of the ISCED (lower secondary education). Along with Austria, on these countries there are no illiterate centenarians. The higher levels of education in Finland are to be expected since the practices of mass schooling were introduced by the German and Scandinavian states (Ramirez & Boli, 1987). In the other countries there is greater heterogeneity, with centenarians distributed across almost all educational levels, although a higher percentage is observed in no formal education (Cyprus), primary education (Lithuania, France, Italy, Romania and Austria) and lower secondary education (Hungary, Croatia, Latvia and Slovenia).

Educational attainment tends to be set early in the life course and reflects social and economic resources at younger ages, but also the political and cultural aspects of each country (Fredman & Lyons, 2012). Data on education and survival from the Survey of Health, Ageing and Retirement in Europe (SHARE)[1] suggest substantial

---

[1] The Survey of Health, Ageing and Retirement in Europe (SHARE) is a multidisciplinary and cross-national panel database of micro data on health, socio-economic status and social and family networks of about 140,000 individuals aged 50 or older (around 380,000 interviews) from 27

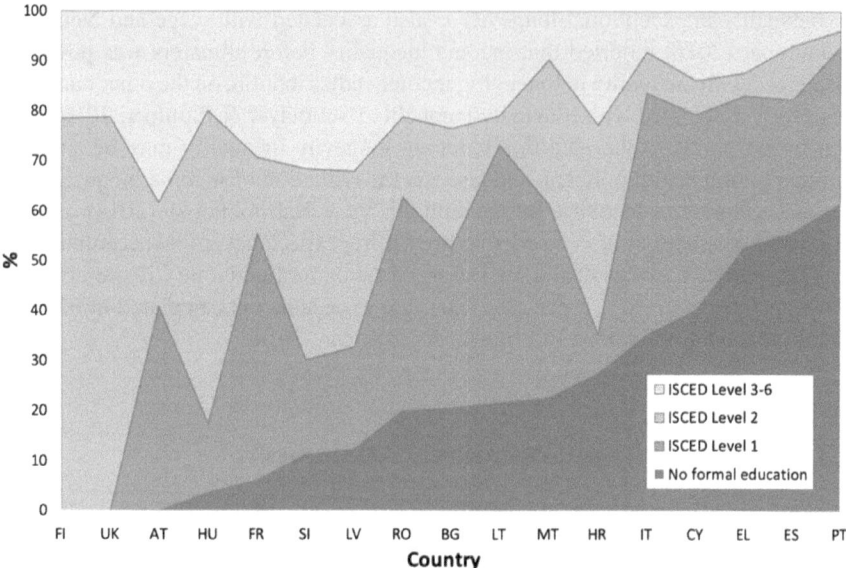

Fig. 3.4 Percentage of centenarians according education level by EU country (*Source* CENSUS HUB | last update: information not available | reference year: 2011 | download date: 13 January 2020)

heterogeneity across Europe. Although lower education attainment was associated with higher mortality rates, this education gradient was relatively small in Southern countries (Bohacek, Crespo, Mira, & Pijoan-Mas, 2015). In the Mediterranean zone, the oldest-old generation still grew up in a time when schooling was not mandatory in some countries and participation in higher education remains everywhere an elite thing up to the cohorts born in the 60s (Ballarino, Meschi, & Scervini, 2013). Fortunately, social norms have changed and advanced schooling has become more attainable, whereby this gap is expected to decrease in future generations.

The current evidence suggests that education, along with an indirect effect through giving access to better jobs and higher incomes, may have direct influences on both health and duration of life (Olshansky et al., 2012). Mirowsky (2003) found that education is more important than income and wealth in connecting better health with higher social status. Well-educated persons have better and more effective habits and attitudes such as personal agency, dependability, good judgment, motivation, effort, trust, and confidence, as well as skills and abilities; they also have more stable and supportive interpersonal relationships (Mirowsky, 2003).

The lower levels of education that generally characterize centenarians should also be particularly considered for research and intervention issues. Researches from international studies on centenarians have been drawing attention to the importance

European countries and Israel. More information on the SHARE project can be found at http://www.share-project.org/home0.html.

Table 3.4 Centenarians according education level (ISCED levels) by EU country

Country	No formal education		ISCED level 1		ISCED level 2		ISCED level 3–6	
	n	%	n	%	n	%	n	%
Austria	0	0.00	454	40.83	232	20.86	426	38.31
Belgium	NA	–	NA	–	NA	–	NA	–
Bulgaria	49	20.59	76	31.93	57	23.95	56	23.53
Croatia	53	27.46	15	7.77	81	41.97	44	22.80
Cyprus	35	40.23	34	39.08	6	6.90	12	13.79
Czechia	NA	–	NA	–	NA	–	NA	–
Denmark	NA	–	NA	–	NA	–	NA	–
Estonia	NA	–	NA	–	NA	–	NA	–
Finland	0	0.00	0	0.00	488	78.46	134	21.54
France	1,108	6.09	8,963	49.27	2,785	15.31	5,336	29.33
Germany	NA	–	NA	–	NA	–	NA	–
Greece	1,311	52.69	756	30.39	115	4.62	306	12.30
Hungary	38	3.64	144	13.79	653	62.55	209	20.02
Ireland	NA	–	NA	–	NA	–	NA	–
Italy	5,410	35.28	7,413	48.34	1,263	8.24	1250	8.15
Latvia	20	12.35	33	20.37	57	35.19	52	32.10
Lithuania	89	21.60	213	51.70	24	5.83	86	20.87
Luxembourg	NA	–	NA	–	NA	–	NA	–
Malta	7	22.58	12	38.71	9	29.03	3	9.68
Netherlands	NA	–	NA	–	NA	–	NA	–
Poland	NA	–	NA	–	NA	–	NA	–
Portugal	940	61.60	478	31.32	49	3.21	59	3.87
Romania	162	19.95	347	42.73	131	16.13	172	21.18
Slovakia	NA	–	NA	–	NA	–	NA	–
Slovenia	25	11.26	42	18.92	85	38.29	70	31.53
Spain	4,560	55.71	2,175	26.57	880	10.75	570	6.96
Sweden	NA	–	NA	–	NA	–	NA	–
United Kingdom	0	0.00	0	0.00	9,655	78.69	2,615	21.31

NA Data not available

Source CENSUS HUB | last update: information not available | reference year: 2011 | download date: 13 January 2020

of methodological strategies in assessing centenarians (Arosio et al., 2017; Jopp & Rott, 2006; Poon et al., 2010). Along with fatigue and compromised sensory and cognitive functions, the level of literacy also plays a significant role. In this sense, especially in countries where centenarians present lower education levels, issues of reliability and validity are to be paid particular attention. Lower education influences the ability of centenarians to take part in social, economic and political activities, as well as to access services and obtain information (e.g., health literacy related), reason why this aspect cannot be devalued.

3.4 Urbanized Centenarians?

The comparison between rural and urban areas allows knowing important aspects related with centenarians' living conditions. Figure 3.5 and Table 3.5 present information on the proportion of centenarians that lived in urban areas by EU country.

For all countries, the proportion of centenarians that lived in such areas is higher than the proportion of centenarians that lived in rural areas (all proportions are higher than 0.5). This is a trend of aging in the EU, since there has been a faster growth of the older population in urban areas compared to rural areas, which can be related to the urbanization of the population across all age groups (United Nations, 2015).

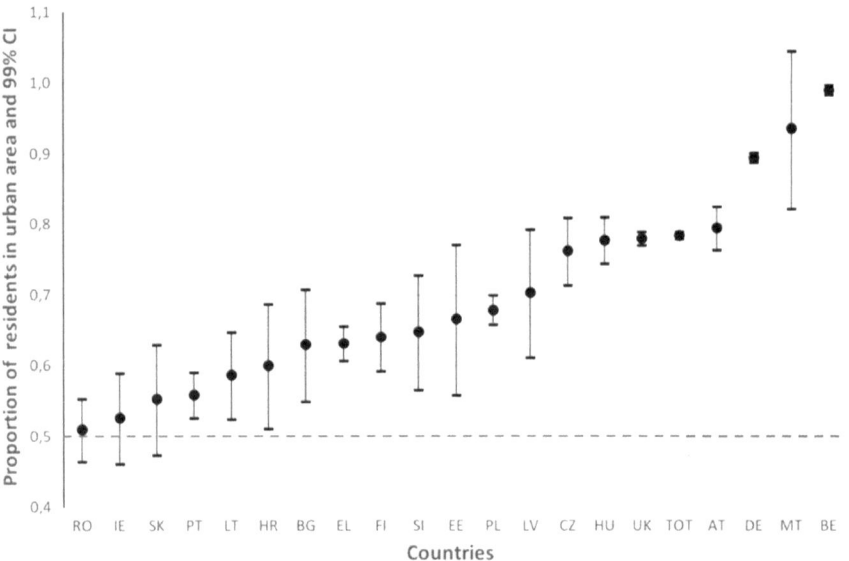

Fig. 3.5 Proportion of centenarians that lived in urban area and 99% confidence interval (CI) by European country (*Source* UNdata I last update: 23 August 2019 I reference year: 2011 I download date: 13 January 2020)

Table 3.5 Centenarians according area of residence (rural/urban) by European country

Country	Rural		Urban	
	n	%	n	%
Austria	228	20.50	884	79.50
Belgium	17	1.04	1,613	98.96
Bulgaria	88	36.97	150	63.03
Croatia	79	39.90	119	60.10
Cyprus	NA	–	NA	–
Czechia	122	23.74	392	76.26
Denmark	NA	–	NA	–
Estonia	42	33.33	84	66.67
Finland	228	35.85	408	64.15
France	NA	–	NA	–
Germany	1,419	10.55	12,026	89.45
Greece	916	36.82	1,572	63.18
Hungary	232	22.22	812	77.78
Ireland	184	47.30	205	52.70
Italy	NA	–	NA	–
Latvia	48	29.63	114	70.37
Lithuania	170	41.26	242	58.74
Luxembourg	NA	–	NA	–
Malta	2	6.45	29	93.55
Netherlands	NA	–	NA	–
Poland	1,046	32.08	2,215	67.92
Portugal	673	44.10	853	55.90
Romania	398	48.95	415	51.05
Slovakia	117	44.66	145	55.34
Slovenia	78	35.14	144	64.86
Spain	NA	–	NA	–
Sweden	NA	–	NA	–
United Kingdom	2,702	22.02	9,566	77.98
Total	8,789	21.55	31,988	78.45

NA Data not available

Source UNdata | last update: 23 August 2019 | reference year: 2011 | download date: 13 January 2020

However, in Romania, Ireland, and Slovakia the proportion of rural and urban residents is similar (i.e., differences between rural and urban proportions are not significant). Perhaps in these countries there is a lower rural-urban split. The countries with higher proportion of centenarians living in urban areas are Belgium (99.0%), Malta (93.5%) and Germany (89.4%). On the other hand, in countries like Lithuania and Portugal, more than 40% of centenarians live in rural areas.

Researchers who have focused on the differences between older adults who live in predominantly rural versus predominantly urban areas have reported several contradictory findings. But it seems to be consensual the existence of positive and negative factors in each type of area, such as, in a negative way, low levels of schooling, poor geographical accessibility to health care services, and high alcohol consumption in rural areas and, on the other hand, poorer air quality, high crime rate and high duration of commuting in urbanized areas (Santana, 2015). Also the lower living standard among rural areas suggests the existence of a higher risk of poverty, posing a possible disadvantage of the rural context in comparison with the urban one (European Comission, 2008).

Specifically on centenarians, available data concerning differences between rural and urban individuals is still very scarce. Back in the early nineties, Clayton et al. (1994) found no differences regarding physical health (both self-reported and objectively measured with physical examination) and mental health, though they found less frequent visits to physicians, less medication use, and less reliance on physical aides (e.g., walkers, hearing aids) among the rural participants (Clayton et al., 1994). More recently, Brandão et al. (2019) showed a better health condition with lower levels of physical and mental impairments, and a greater number of centenarians who did not succumb to the three most common lethal diseases (heart disease, non-skin cancer and stroke) in the rural areas (Brandão, Ribeiro, Afonso, & Pául, 2019).

The variance across studies and the proportions of rural/urban centenarians (Fig. 3.5) can also be related with the difficulties in defining rural and urban areas. In Europe, individual countries have very different official definitions. For instance, the definition of urban/rural areas may be established considering a combination of the following criteria: size of population, population density, distance between built-up areas, predominant type of economic activity, conformity to legal or administrative status, and urban characteristics such as specific services and facilities (United Nations, 2019a). Rural areas are characterized by dispersed population and agricultural-based economy, resulting in lack of access to major services (European Comission, 2008). The pattern of centenarians presented is representative of general population distribution and the territorial organization of the country, with countries in Eastern and Southern Europe and Ireland presenting predominance of rural regions (European Comission, 2008).

3.5   A Population of Expected Widows

Considering the family life of centenarians, and bearing in mind the previously exposed sex gap in human longevity, as it would be expected, very few centenarians reach this age with a living spouse. Therefore, widowhood is the modal marital status across studies. As it can be observed in Fig. 3.6, in all EU countries, the majority of centenarians are widowed, with percentages varying between 63.51% (Hungary) and 91.88% (Croatia). Ireland is the country with the highest percentage of single centenarians (i.e., that have never been married), reaching up to 20.57%, followed by Malta (19.35%) and Spain (16.79%); the lower percentages are in Croatia (2.54%), Greece (3.58%) and Bulgaria (5.04%) (Table 3.6).

As the incidence of widowhood increases with age, these higher rates are to be expected. But one particularly aspect of centenarians' old widowhood is the long periods of absence of a partner relationship, since many of them may have lost their spouse a long time ago (MacDonald, 2007). In the cases of widowhood and divorce, women present higher odds of not remarrying than do men, and older persons have a smaller chance of remarrying than do younger people (Bumpass, Sweet, & Martin, 1990).

Although it is not so high as the widow category, there are as also more single female centenarians (Fig. 3.7a). Hence, as expected, the proportion of married men is much higher (Fig. 3.7b). Some countries present great differences across sex in single and married status. The cases that draw the most attention are Ireland, with the higher percentage of single males, and Malta, with the higher percentage of single

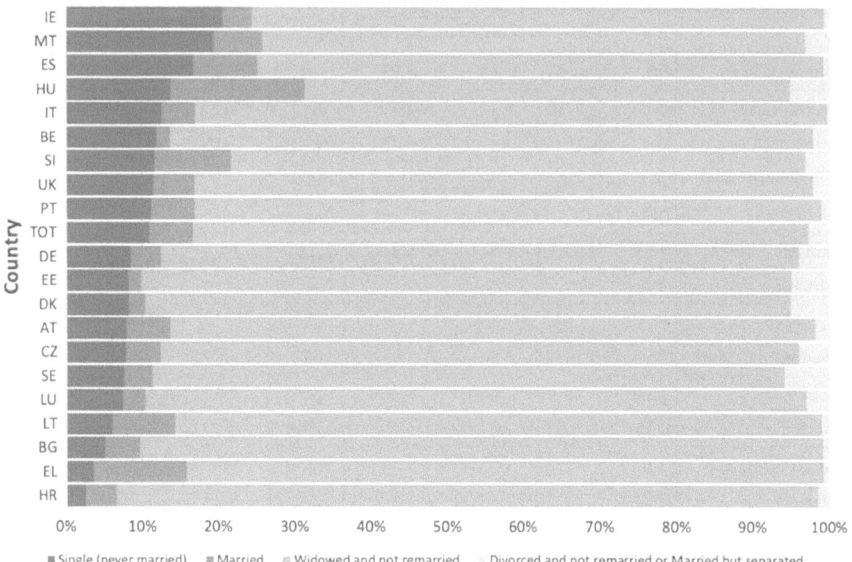

Fig. 3.6 Proportion of centenarians according marital status by EU country (*Source* UNdata ǀ last update: 23 August 2019 ǀ reference year: 2011 ǀ download date: 14 January 2020)

Table 3.6 Centenarians according marital status by EU country

Country	Marital status							
	Single (never married)		Married		Widowed and not remarried		Divorced and not remarried or married but separated	
	n	%	n	%	n	%	n	%
Austria	88	7.91	65	5.85	938	84.35	21	1.89
Belgium	195	11.97	29	1.78	1,369	84.04	36	2.21
Bulgaria	12	5.04	11	4.62	213	89.50	2	0.84
Croatia	5	2.54	8	4.06	181	91.88	3	1.52
Cyprus	NA	–	NA	–	NA	–	NA	–
Czechia	40	7.89	23	4.54	424	83.63	20	3.94
Denmark	74	8.22	20	2.22	761	84.56	45	5.00
Estonia	10	8.26	2	1.65	103	85.12	6	4.96
Finland	NA	–	NA	–	NA	–	NA	–
France	NA	–	NA	–	NA	–	NA	–
Germany	1,144	8.51	542	4.03	11,223	83.49	533	3.97
Greece	89	3.58	305	12.26	2,075	83.40	19	0.76
Hungary	144	13.79	183	17.53	663	63.51	54	5.17
Ireland	80	20.57	15	3.86	291	74.81	3	0.77
Italy	1,909	12.66	654	4.34	12,476	82.74	40	0.27
Latvia	NA	–	NA	–	NA	–	NA	–
Lithuania	25	6.07	34	8.25	349	84.71	4	0.97
Luxembourg	5	7.46	2	2.99	58	86.57	2	2.99
Malta	6	19.35	2	6.45	22	70.97	1	3.23
Netherlands	NA	–	NA	–	NA	–	NA	–
Poland	NA	–	NA	–	NA	–	NA	–
Portugal	172	11.27	87	5.70	1,251	81.98	16	1.05
Romania	NA	–	NA	–	NA	–	NA	–
Slovakia	NA	–	NA	–	NA	–	NA	–
Slovenia	26	11.71	22	9.91	167	75.23	7	3.15
Spain	1,375	16.79	680	8.30	6,070	74.11	65	0.79
Sweden	135	7.63	67	3.79	1,464	82.71	104	5.88
United Kingdom	1,419	11.57	651	5.31	9,930	80.94	268	2.18
Total	7,101	11.28	3,631	5.52	52,352	81.17	1,269	2.03

NA Data not available

Source UNdata I last update: 23 August 2019 I reference year: 2011 I download date: 14 January 2020

a)

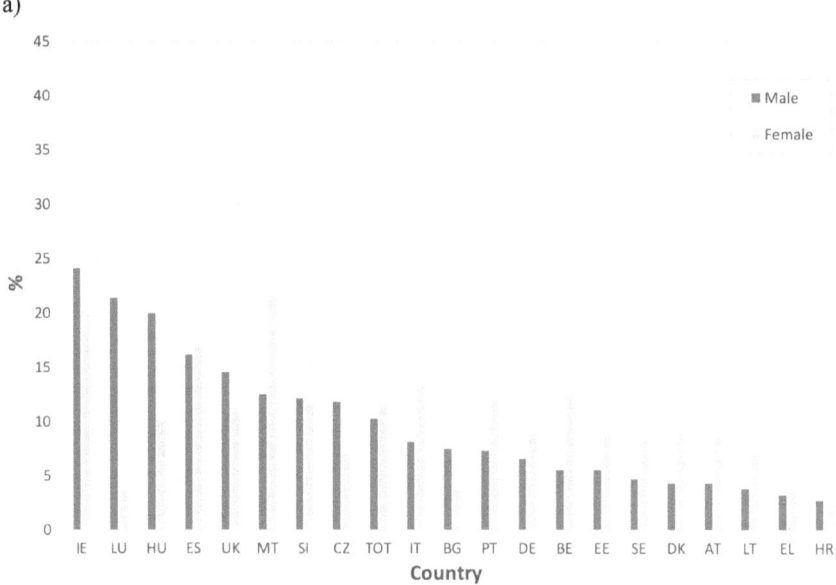

b)

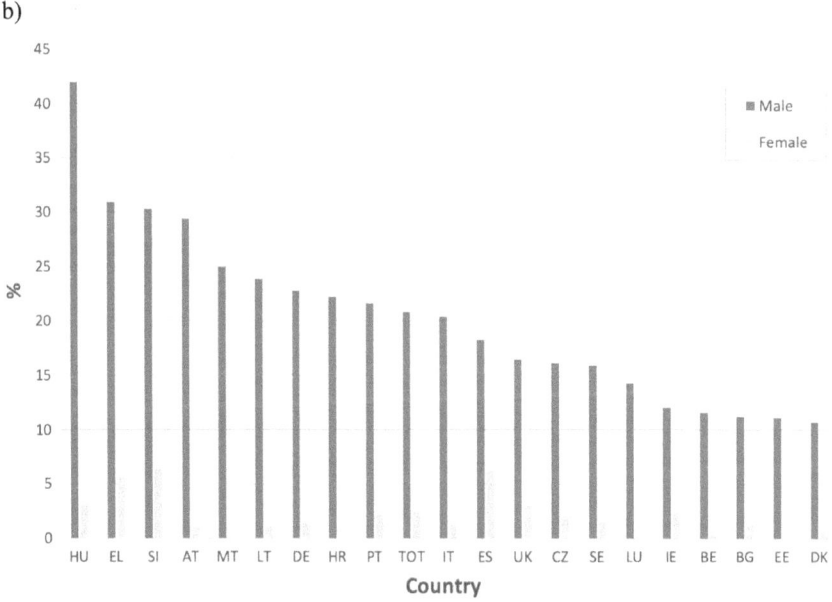

Fig. 3.7 Proportion of centenarians according marital status and sex by EU country: **a)** Single (never married) and **b)** Married (*Source* UNdata | last update: 23 August 2019 | reference year: 2011 | download date: 14 January 2020)

females (Table 3.7). Also worthwhile noting is the exceptionally high percentage (above 30%) of married male centenarians in Hungary, Greece and Slovenia (while the TOT is 20.84%).

The type of marital status, depending on other social and cultural variables, can be reflected in living arrangements, as it will be next presented.

Table 3.7 Centenarians according marital status (single and married) and sex by EU country

Country	Marital status							
	Single (never married)				Married			
	Female		Male		Female		Male	
	n	%	n	%	n	%	n	%
Austria	80	8.65	8	4.28	10	1.08	55	29.41
Belgium	184	12.86	11	5.56	6	0.42	23	11.62
Bulgaria	6	3.80	6	7.50	2	1.27	9	11.25
Croatia	4	2.48	1	2.78	0	0.00	8	22.22
Cyprus	NA	–	NA	–	NA	–	NA	–
Czechia	29	7.00	11	11.83	8	1.93	15	16.13
Denmark	68	8.95	6	4.29	5	0.66	15	10.71
Estonia	9	8.74	1	5.56	0	0.00	2	11.11
Finland	NA	–	NA	–	NA	–	NA	–
France	NA	–	NA	–	NA	–	NA	–
Germany	1,033	8.78	111	6.62	160	1.36	382	22.79
Greece	68	3.70	21	3.23	104	5.66	201	30.92
Hungary	67	10.18	77	19.95	21	3.19	162	41.97
Ireland	66	19.94	14	24.14	8	2.42	7	12.07
Italy	1,709	13.54	200	8.13	153	1.21	501	20.37
Latvia	NA	–	NA	–	NA	–	NA	–
Lithuania	20	7.09	5	3.85	3	1.06	31	23.85
Luxembourg	2	3.77	3	21.43	0	0.00	2	14.29
Malta	5	21.74	1	12.50	0	0.00	2	25.00
Netherlands	NA	–	NA	–	NA	–	NA	–
Poland	NA	–	NA	–	NA	–	NA	–
Portugal	152	12.13	20	7.33	28	2.23	59	21.61
Romania	NA	–	NA	–	NA	–	NA	–
Slovakia	NA	–	NA	–	NA	–	NA	–
Slovenia	22	11.64	4	12.12	12	6.35	10	30.30

(continued)

Table 3.7 (continued)

Country	Marital status							
	Single (never married)				Married			
	Female		Male		Female		Male	
	n	%	n	%	n	%	n	%
Spain	1,140	16.91	235	16.21	415	6.16	265	18.28
Sweden	122	8.17	13	4.71	23	1.54	44	15.94
United Kingdom	1117	10.96	302	14.55	310	3.04	341	16.43
Total	5,903	11.49	1,050	10.25	1,268	2.47	2,134	20.84

NA Data not available
Source UNdata | last update: 23 August 2019 | reference year: 2011 | download date: 14 January 2020

3.6 Living Mostly at Home

Figure 3.8 presents information regarding the proportion of centenarians that lived in a set of collective living quarters, which includes institutions and care facilities, by EU country and by sex. As it can be seen, in the considered countries 32.57% of centenarians lived in institutions, meaning that the majority of centenarians lived in the community. Germany, Ireland, Belgium, Malta and Luxembourg are the exceptions presenting a higher percentage of centenarians who were institutionalized: 50.74%, 52.96%, 54.05%, 54.84% and 55.22%, respectively. Romania, Greece and

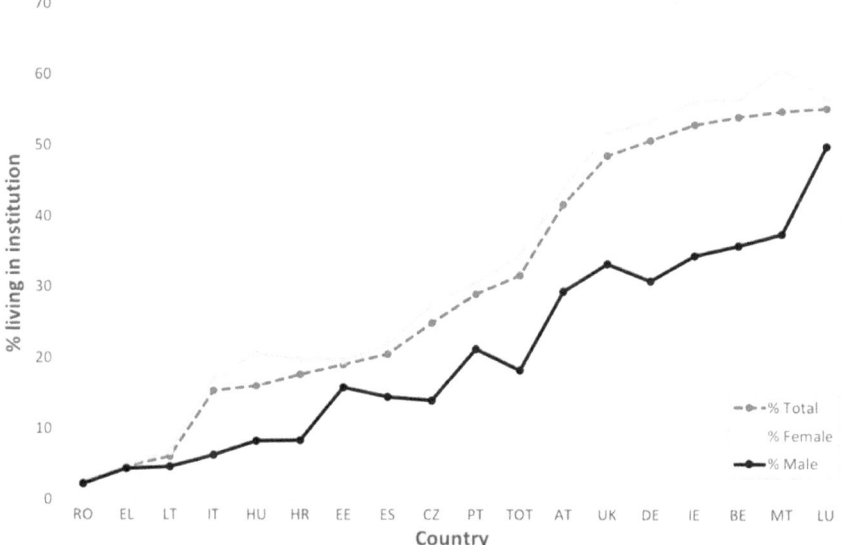

Fig. 3.8 Proportion of centenarians that lived in institution by sex and by EU country (*Source* UNdata | last update: 23 August 2019 | reference year: 2011 | download date: 14 January 2020)

Lithuania, on the other hand, are the countries with lower percentages of centenarians living in an institution, with values varying between 2.46% (Romania) and 6.07% (Lithuania). A mention is to be made to the three centenarians who were reported as being homeless: one man in Czechia, one woman in Hungary, and one man in Italy.

Meaningful life changes that frequently occur in older adulthood (e.g., widowhood, change in economic status, health decline) may result in changes in living situations (Fredman & Lyons, 2012). And there are striking gender differences in living arrangement among older adults. When analyzing differences between centenarian men and women (Table 3.8), the percentage of men that lived in institutions is lower when compared with that of women, in all countries. A significant association between sex and living arrangement was found for the following countries: Austria (p < 0.001), Belgium (p < 0.001), Czechia (p = 0.008), Germany (p < 0.001), Hungary (p < 0.001), Ireland (p = 0.004), Italy (p < 0.001), Portugal (p = 0.002), Spain (p < 0.001) and the United Kingdom (p < 0.001).

Regardless of the great effort of the European policy to support "ageing in place" policies and assure home care services that acknowledge older adults' preferences of staying at their own houses for as long as possible, there are still differences affecting older people's ability to age in their communities across countries. This may be strongly related with housing conditions, income levels, place of residence, the availability of care and support services, physical environment, and one's health and physical capacities (Mestheneos, 2011). In relation to centenarians, the impact of these factors can be even more expressive due to the augmented diversity of their health and social care needs. Indeed, centenarian studies have been showing that very old adults have somehow distinctive health problems and care needs that may demand more specific services (Rasmussen & Andersen-Ranberg, 2016).

Living in the community or in an institution may also reflect the national policies and welfare state services and family support for the elderly as well as cultural characteristics related to the desire to maintain a sense of autonomy (Hank, 2007). Income difficulties, unsuitable and inaccessible housing (Serra, Watson, Sinclair, & Kneale, 2011) or lack of informal caregiving have been reported as playing an important role in the institutionalization process of centenarians (Ribeiro & Araújo, 2016). Although referring to the American context, also Randall et al. (2010), within the Georgia Centenarian Study, found that centenarians' living arrangements were associated with social resources, in the sense that having a greater network contact reduced the odds of centenarians residing in a nursing home (Randall et al., 2010).

Centenarians who were not represented in the previous data are the ones who live in the community, mainly with their relatives or by themselves/ alone. Figure 3.9 provides information about centenarians that lived alone by sex and by EU country. The EU average proportion (TOT) of one-person households is of 44.67%. Nevertheless, there is a high heterogeneity across countries. Lithuania is the country with lower percentage of centenarians living alone (4.65%), followed by Spain (18.99%) and Portugal (20.02%). But in countries like Romania and the United Kingdom more that than 60% of the centenarians live alone, 62.32% and 61.98%, respectively. This high proportion observed in the UK is quite expectable since in the second half of the twentieth century there has been a high increase in the proportion of older adults living alone (Falkingham, Evandrou, McGowan, Bell, & Bowes, 2010).

Table 3.8 Centenarians living in collective living quarters by gender and by EU country

Country	Collective living quarters					
	Total		Female		Male	
	n	%	n	%	n	%
Austria	463	41.64	408	44.11	55	29.41
Belgium	881	54.05	810	56.56	71	35.86
Bulgaria	NA	–	NA	–	NA	–
Croatia	35	17.68	32	19.75	3	8.33
Cyprus	NA	–	NA	–	NA	–
Czechia	128	24.90	114	27.54	14	14.00
Denmark	NA	–	NA	–	NA	–
Estonia	24	19.05	21	19.63	3	15.79
Finland	NA	–	NA	–	NA	–
France	NA	–	NA	–	NA	–
Germany	6,768	50.74	6,255	53.56	513	30.90
Greece	116	4.74	88	4.86	28	4.39
Hungary	167	16.00	135	20.52	32	8.29
Ireland	206	52.96	186	56.19	20	34.48
Italy	2,310	15.32	2,156	17.09	154	6.26
Latvia	NA	–	NA	–	NA	–
Lithuania	25	6.07	19	6.74	6	4.62
Luxembourg	37	55.22	30	56.60	7	50.00
Malta	17	54.84	14	60.87	3	37.50
Netherlands	NA	–	NA	–	NA	–
Poland	NA	–	NA	–	NA	–
Portugal	442	28.96	384	30.65	58	21.25
Romania	20	2.46	14	2.56	6	2.26
Slovakia	NA	–	NA	–	NA	–
Slovenia	NA	–	NA	–	NA	–
Spain	1,680	20.53	1,470	21.83	210	14.48
Sweden	NA	–	NA	–	NA	–
United Kingdom	5,956	48.55	5,265	51.65	691	33.30
Total	19,275	32.57	17,401	35.35	1,874	18.82

NA Data not available

Source UNdata | last update: 23 August 2019 | reference year: 2011 | download date: 14 January 2020

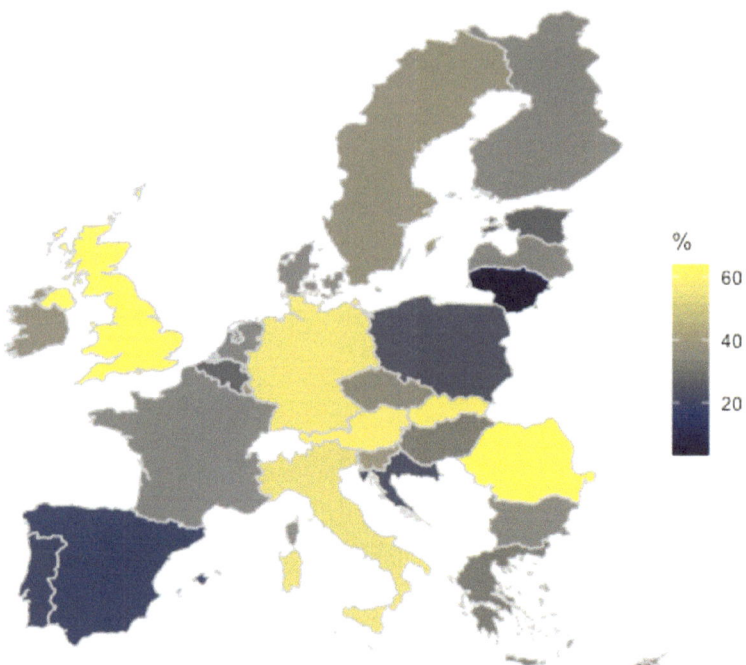

Fig. 3.9 Proportion of centenarians living alone by EU country (*Source* UNdata | last update: 23 August 2019 | reference year: 2011 | download date: 14 January 2020)

When analyzing the percentages by sex (Table 3.9), we can see that, in overall, women have higher odds of living alone, with exception of Croatia, Estonia, Luxembourg, Poland and Portugal. In Malta no centenarian male was living alone.

Previous studies about co-residence across European countries point to North–South differences, with a low rate of child/elderly cohabitation in the Nordic countries. On the other hand, in Southern countries like Greece, Italy, Spain, but also in Poland (that are considered as "strong family countries", i.e., where family bonds are especially tight), there is a greater trend of co-residential familial arrangements (Hank, 2007; Lyberaki, Tinios, Mimis, & Georgiadis, 2013). Also multi-generation households are not very common in Western Europe but much more common in Eastern countries, especially Lithuania, Romania, Bulgaria, Poland and Slovenia (European Comission, 2008).

Living arrangements are of crucial importance as determinants of older adults' financial and social situation, as well as of their wellbeing and sense of loneliness (Jong-Gierveld, Valk, & Blommesteijn, 2011); particularly living alone has been associated with increased risk of depression, worse physical health outcomes and nursing home placement (Fredman & Lyons, 2012). But the cases of centenarians who live with their families, normally children, should also be focus of attention, as children of centenarians may themselves be aged and, therefore, have themselves an extensive range of care needs (Serra et al., 2011). As Boerner, Jopp, Park, and Rott (2016, p. 165) have stated, "supportive programs and policies that were designed with younger older adults in mind do not necessarily meet the needs of very old adults and their families", whereby housing and caring of very old and centenarians can assume very specific situations in this advanced stage of life and are important subjects that matter to address (Boerner et al., 2016).

Table 3.9 Centenarians living alone by sex and by European country

Country	One-person household					
	Total		Female		Male	
	n	%	n	%	n	%
Austria	378	58.24	320	61.90	58	43.94
Belgium	NA	–	NA	–	NA	–
Bulgaria	NA	–	NA	–	NA	–
Croatia	36	22.09	28	21.54	8	24.24
Cyprus	NA	–	NA	–	NA	–
Czechia	137	35.40	107	35.55	30	34.88
Denmark	NA	–	NA	–	NA	–
Estonia	28	27.18	22	25.29	6	37.50
Finland	NA	–	NA	–	NA	–
France	NA	–	NA	–	NA	–
Germany	3,666	56.39	3,160	58.88	506	44.62
Greece	736	31.57	550	31.96	186	30.49
Hungary	283	32.31	187	35.82	96	27.12
Ireland	70	37.84	57	38.78	13	34.21
Italy	6,980	54.15	5,783	54.73	1,197	27.85
Latvia	NA	–	NA	–	NA	–
Lithuania	18	4.65	15	5.70	3	2.42
Luxembourg	12	40.00	9	39.13	3	42.86
Malta	4	28.57	4	44.44	0	0.00
Netherlands	NA	–	NA	–	NA	–
Poland	619	23.63	470	22.71	149	27.14
Portugal	217	20.02	168	19.33	49	22.79
Romania	NA	–	NA	–	NA	–
Slovakia	NA	–	NA	–	NA	–
Slovenia	NA	–	NA	–	NA	–
Spain	1,235	18.99	1,030	19.56	205	16.53
Sweden	NA	–	NA	–	NA	–
United Kingdom	3,912	61.98	3,206	65.06	706	51.01
Total	18,331	44.67	15,116	46.11	3,215	38.97

NA Data not available

Source UNdata | last update: 23 August 2019 | reference year: 2011 | download date: 14 January 2020

3.7   The Impact of Cardiovascular Diseases in Mortality

The increasing number of centenarians has allowed the use of diverse approaches to gain insights into the health, disease patterns, and causes and places of death of the oldest old. Most common reported causes of death among centenarians regard respiratory diseases, circulatory diseases, and pneumonia (e.g., Berzlanovich et al., 2005; Chen et al., 2019; Yu, Tam, & Woo, 2017).

Although not being exclusively focused on centenarians due to limitations related to the sources here considered (WHO Mortality Database) and/or to the fact that they do not cover a wide range of countries due to unavailable data, Table 3.10 and Fig. 3.10 provide information on the main causes of death (ICD-10 classification) of people aged 95+ years old by sex and by EU country. In overall, we can observe the presence of five clusters of countries presenting the same leading cause of

Table 3.10 Percentage of main cause of death of people with 95+ years by sex and by EU country

Country	Female				Male			
	No. of people with 95+ years	Main cause of death	n	%	No. of people with 95+ years	Main cause of death	n	%
Austria	8,707	Atherosclerotic heart disease	523	6.01	202	Atherosclerotic heart disease	115	56.93
Belgium	12,412	Heart failure, unspecified	303	2.44	2,610	Heart failure, unspecified	81	3.1
Bulgaria	NA	NA	NA	–	NA	NA	NA	–
Croatia	NA	Ischaemic cardiomyopathy	97	–	NA	Atherosclerotic heart disease	23	–
Cyprus	424	Heart failure, unspecified	19	4.48	269	Stroke. not specified as haemorrhage or infarction	9	3.35
Czechia	4,709	Chronic ischaemic heart disease, unspecified	447	9.49	1,208	Chronic ischaemic heart disease, unspecified	112	9.27
Denmark	6,688	Senility	263	3.93	1,635	Senility	50	3.06
Estonia	653	Atherosclerotic heart disease	60	9.19	188	Atherosclerotic heart disease	22	11.70
Finland	NA	NA	NA	–	NA	NA	NA	–
France	NA	Heart failure, unspecified	1983	–	NA	Heart failure, unspecified	494	–
Germany	130,124	Heart failure, unspecified	4,574	3.52	58,418	Heart failure, unspecified	818	1.40
Greece	NA	NA	NA	–	NA	NA	NA	–

(continued)

Table 3.10 (continued)

Country	Female				Male			
	No. of people with 95+ years	Main cause of death	n	%	No. of people with 95+ years	Main cause of death	n	%
Hungary	NA	Chronic ischaemic heart disease, unspecified	395	–	NA	Chronic ischaemic heart disease, unspecified	110	–
Ireland	NA	Acute myocardial infarction, unspecified	95	–	NA	Acute myocardial infarction, unspecified	27	–
Italy	NA	Chronic ischaemic heart disease, unspecified	2,160	–	NA	Chronic ischaemic heart disease, unspecified	659	–
Latvia	1,284	Atherosclerotic heart disease	153	11.92	246	Atherosclerotic heart disease	40	16.26
Lithuania	2,009	Atherosclerotic heart disease	202	10.05	575	Atherosclerotic heart disease	50	8.70
Luxembourg	367	Heart failure, unspecified	23	6.27	54	Heart failure, unspecified	4	7.41
Malta	239	Unspecified dementia	11	4.60	87	Chronic ischaemic heart disease, unspecified	5	5.75
Netherlands	15,573	Unspecified dementia	702	4.51	3,290	Pneumonia, unspecified	120	3.65
Poland	NA	Senility	1719	–	NA	Senility	453	–
Portugal	NA	NA	NA	–	NA	NA	NA	–
Romania	NA	Chronic ischaemic heart disease, unspecified	477	–	NA	Chronic ischaemic heart disease, unspecified	237	–
Slovakia	NA	NA	NA	–	NA	NA	NA	–
Slovenia	NA	NA	NA	–	NA	NA	NA	–
Spain	57,323	Unspecified dementia	1,456	2.54	18,983	Heart failure, unspecified	341	1.80
Sweden	13,412	Unspecified dementia	439	3.27	3,630	Heart failure, unspecified	124	3.42
United Kingdom	NA	Senility	3,445	–	NA	Senility	584	–

NA Data not available

Source WHO Morality Database | last update: information not available | reference year: 2011 | download date: 14 January 2020

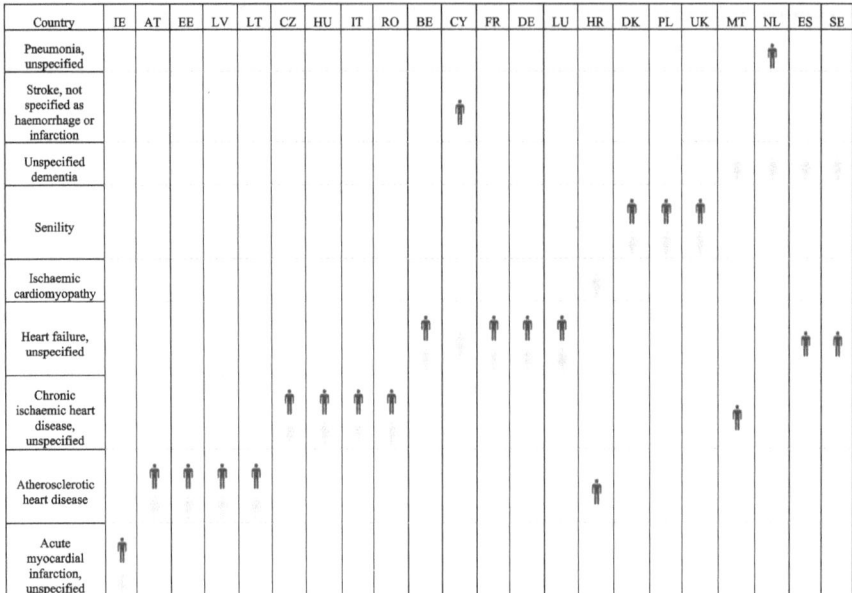

Fig. 3.10 Main cause of death by sex and by EU country *(Source* WHO Morality Database | last update: information not available | reference year: 2011 | download date: 14 January 2020)

death for both sexes: (i) Austria, Estonia, Latvia, and Lithuania—Atherosclerotic heart disease; (ii) Czechia, Hungary, Italy and Romania—Chronic ischemic heart disease, unspecified; (iii) Belgium, France, Germany, and Luxembourg—Heart failure, unspecified; (iv) Denmark, Poland, and United Kingdom—Senility; and (v) Ireland—Acute myocardial infarction, unspecified.

Table 3.11 compiles the most commonly causes of death identified into broader categories of the ICD-10, namely: (i) Diseases of the circulatory system; (ii) Symptoms, signs, and abnormal clinical laboratory findings, not elsewhere classified; (iii) Neoplasms; (iv) Diseases of the respiratory system; (v) Endocrine, nutritional and metabolic diseases; and (vi) Mental and behavioral disorders. As it can be observed, the first group unequivocally leads the causes of deaths in all countries considered, being that most noticeable sex differences can be found in Germany with women more expressively outnumbering men. In overall, this finding is in line with recent available information for German centenarians who present cardiovascular conditions as the only prevalent disease group with high mortality risk, though no gender differences could be made due to gendered survival ratio that resulted in a small number of males (cf. Jopp, Boerner, & Rott, 2016).

The causes of death here presented do not differ much from the ones observed at younger ages. A recent OECD report showed that ischemic heart disease, stroke or other circulatory diseases cause more than one in three deaths, and that one in four deaths are due to cancer (OECD, 2017). Nevertheless, since 1990 the rate of mortality for circulatory disease has fallen, on average, 50%, and cancer mortality

Table 3.11 Percentage of deaths of people with 95+ years according five groups of ICD-10 by sex and by EU country

Country	Diseases of the circulatory system		Symptoms, signs and abnormal clinical and laboratory findings, not elsewhere classified		Neoplasms		Diseases of the respiratory system		Endocrine, nutritional and metabolic diseases		Mental and behavioural disorders	
	Female	Male	Female	Male	Female	Male	Female	Male	Female	Male	Female	Male
Austria	22.98	22.3	2.17	1.83	2.24	4.2	1.45	3.21	1.08	0.74	0.33	0.45
Belgium	12.5	13.68	3.71	3.1	2.14	4.1	3.92	7.36	1.03	0.88	2.2	1.72
Bulgaria	–	–	–	–	–	–	–	–	–	–	–	–
Croatia	–	–	–	–	–	–	–	–	–	–	–	–
Cyprus	16.98	14.5	4.25	4.09	0.24	1.86	2.83	3.35	2.59	0.37	1.42	0.37
Czechia	28.12	27.48	0.68	0.58	1.61	3.73	2.06	4.06	0.66	0.75	0.55	0.58
Denmark	11.32	12.42	7.07	5.57	2.27	4.22	3.84	5.08	0.82	1.04	3.01	2.2
Estonia	36.91	30.85	4.29	2.66	3.06	4.26	0.31	1.6	0	0	0.31	0.53
Finland	–	–	–	–	–	–	–	–	–	–	–	–
France	–	–	–	–	–	–	–	–	–	–	–	–
Germany	16.98	7.3	0.93	0.37	1.56	1.2	1.64	1.19	1.03	0.37	1.66	0.54
Greece	–	–	–	–	–	–	–	–	–	–	–	–
Hungary	–	–	–	–	–	–	–	–	–	–	–	–
Ireland	–	–	–	–	–	–	–	–	–	–	–	–
Italy	–	–	–	–	–	–	–	–	–	–	–	–
Latvia	24.84	27.24	9.03	4.07	1.56	2.85	0.16	0.81	0.31	0.41	0.16	0
Lithuania	33.05	28.7	0.15	0.87	1	1.74	0.45	1.04	0.05	0	0	0

(continued)

Table 3.11 (continued)

Country	Diseases of the circulatory system		Symptoms, signs and abnormal clinical and laboratory findings, not elsewhere classified		Neoplasms		Diseases of the respiratory system		Endocrine, nutritional and metabolic diseases		Mental and behavioural disorders	
	Female	Male	Female	Male	Female	Male	Female	Male	Female	Male	Female	Male
Luxembourg	19.35	31.48	3.54	5.56	2.72	3.7	4.9	14.81	0.54	1.85	1.91	1.85
Malta	17.99	18.39	0.84	0	1.67	3.45	5.02	5.75	0	0	5.44	2.3
Netherlands	11.01	13.19	3.4	3.47	2.27	3.56	3.19	6.44	1.24	1.19	4.91	3.4
Poland	–	–	–	–	–	–	–	–	–	–	–	–
Portugal	–	–	–	–	–	–	–	–	–	–	–	–
Romania	–	–	–	–	–	–	–	–	–	–	–	–
Slovakia	–	–	–	–	–	–	–	–	–	–	–	–
Slovenia	–	–	–	–	–	–	–	–	–	–	–	–
Spain	13.35	10.75	1.75	1.47	1.85	3.48	4.1	6.29	1.07	0.81	3.01	1.79
Sweden	17.52	19.42	3.95	2.89	1.51	3.44	1.92	3.22	0.76	0.96	3.88	2.87
United Kingdom	–	–	–	–	–	–	–	–	–	–	–	–

Source WHO Morality Database I last update: information not available I reference year: 2011 I download date: 14 January 2020

18%. So it seems that many of these centenarians in fact delay and/or smooth the most common diseases. Relevant may be the occurrence of multi-morbidity that rise with age and varies with socioeconomic context (OECD, 2011).

Chapter 4
Centenarian Studies Across Europe

Throughout the world the number of centenarian studies has been increasing. Once very limited in number and in their infancy, they have entered into a maturity stage over the last decades (Willcox, Willcox, & Poon, 2010), currently providing an increasingly robust repository of information on demographic, genetic, biological, psychosocial and clinical data from which researchers, health care professionals, social agents and policy makers can rely on. In fact, ever since the establishment of the world's longest centenarian study in 1975, the Okinawa Centenarian Study (Craig Willcox, Willcox, Hsueh, & Suzuki, 2006), and of the longest ongoing centenarian study in the USA, the Georgia Centenarian Study (Poon et al., 1992), that several other studies have been launched worldwide—some for more than a decade, others established more recently as an echo of clear demographic insights on the increasing number of individuals reaching 100 years old. As a matter of fact, the list of developed countries presenting their own centenarian study not only has increased, as the topic of extreme longevity has gained greater limelight, making the study of centenarians a preeminent topic in several scientific international meetings.

Particularly within Europe, several groups of researchers have been conducting studies with centenarians and near centenarians. An example of a long-standing centenarian study is the Longitudinal Study of Danish Centenarians (Andersen-Ranberg & Jeune, 2006), but in a search of the literature using the PubMed database conducted in 2012, along with Denmark at least 16 more European countries were identified as conducting oldest-old or centenarian research (Poon & Cheung, 2012). These were Azerbaijan, Belgium, Finland, France, Georgia, Germany, Greece, Hungary, the Netherlands, Poland, Italy, Russia, Spain, Sweden, Switzerland, and the United Kingdom. All these studies differ in several ways, namely in their focus and objectives (e.g., genetic vs. psychosocial vs. multidisciplinary), samples (population-based vs. convenient samples) and methodological approach (e.g., qualitative vs. quantitative; secondary analysis of administrative data vs. primary data). These differences have made it difficult to compare findings and even more difficult to establish cross-country feasible comparisons. Nevertheless, some groups of researchers from

L. Teixeira et al., *Centenarians*, SpringerBriefs in Aging,
https://doi.org/10.1007/978-3-030-52090-8_4

different countries have attempted to merge information to create multicenter and multicultural studies, allowing for an improvement in the sample size and a comparison of different realities. The Five Country Oldest Old Project (5-COOP)[1] and the International Centenarian Consortium (ICC)[2] are two successful and illustrative examples.

4.1 Ongoing International Collaboration Initiatives

The Five Country Oldest Old Project (5-COOP) aims to accurately compare the health status of centenarians living in five countries—Denmark, France, Japan, Sweden, and Switzerland, all of which present high-quality data on mortality and/or health for centenarians, and include a high number of centenarians in their populations. This project focuses on the cohorts of people born in 1911 (and later) who reached their 100th birthday from 2011 onwards (Robine et al., 2010), and aims to examine a diversity of factors related to longevity as physical, sensory and cognitive capacities, and geriatric conditions. Along with forecasting the demographics of the oldest old, the project also intends to increase the understanding of the causes and patterns of diseases, estimate the risk of dependence and physical or cognitive impairment, and the health and social needs of this population. To date, several publications have been made available on a diverse range of topics, including the centenarians' overall characteristics of the five considered countries (Herrmann & Zekry, 2017), information on frailty profiles (Herr et al., 2018), use of healthcare services and assistive devices (Dupraz et al., 2020), and levels of mortality selection (Robine et al., 2010). In overall, by merging five standardized datasets, the 5-COOP project allowed an important international step in centenarian research with a transnational focus (Robine, 2019), increasing the available knowledge to be used in care resources planning and population forecasts.

The International Centenarian Consortium (ICC) is a group of researchers from around the world (Europe, USA, Japan, Korea, Honk Kong, and Australia) who have a special interest in centenarians. The purpose of the consortium is to share recent information and research on exceptional human longevity and to foster collaboration among different centenarian research teams (e.g., joint research protocols, grant proposal writing and publications) (Iowa State University, 2020). The first ICC meeting, in which the group was officially established, took place in Athens (Georgia, USA) in 1994. Since then meetings have occurred in several European countries (Sweden in 1995, 1998 and 2012; Germany in 1996 and 1999, France in 1996; Italy in 2015; Portugal in 2016; Switzerland in 2019) (Poon, Hagberg, & Martin, 2004),

---

[1] More information on the 5-COOP study can be found at https://www.axa-research.org/en/project/jean-marie-robine.

[2] More information on the ICC can be found at https://research.hs.iastate.edu/international-centenarian-consortium/.

and in 2012, a specific subgroup of researchers interested in studying neurocognitive disorders in the oldest-old formed the International Centenarian Consortium of Dementia (ICC-Dementia). This subgroup comprises eighteen different studies from Europe, Asia, USA, and Oceania, and has as main goals to (i) enable the comparison of data referring to brain ageing; (ii) validate the diagnosis of dementia by applying common diagnostic criteria and examining the rate of cognitive decline in those cohorts using different approaches; (iii) examine and compare cross-nationally risk and protective factors for dementia at the extreme end of life; and (iv) identify factors that predict successful brain ageing into the 11th decade of life that are robust across cohorts (Brodaty et al., 2016; Centre for Healthy Brain Ageing, 2020).

Table 4.1 lists the ICC and the ICC-Dementia studies (Centre for Healthy Brain Ageing, 2020; Iowa State University, 2020). The studies from the ICC have in common the fact that they focus specifically on (near) centenarians but they incorporate distinct interests, approaches and methodologies.

From the country with the highest CR comes the French Centenarian Study, which is conducted by the Ipsen Foundation. It enrolls 910 subjects—800 centenarians, one of them being Jeanne Calment, the oldest human whose age was well-documented (Allard & Robine, 2000). This study had a focus on health and mental status, and functional capacity (Robine, Romieu, & Allard, 2003).

From Italy, the third country with a higher CR, two studies take part of the ICC. The Monzino 80+ Study is a population-based study of residents aged 80 years or older in the Varese province, in which of the 1371 eligible individuals, 64 were centenarians (Lucca et al., 2020). Its main objective is to study dementia incidence and prevalence. Secondly, there is the "Centenari a Trieste (CaT)", a population-based study of centenarians living in the province of Trieste, in which 163 centenarians were identified and 70 participated in the first wave of the study. This study included neuropsychological testing, clinical examinations, blood drawing, and administrative data access (Tettamanti & Marcon, 2018). Also from the Mediterranean region, in Portugal, there is the PT100 Oporto Centenarian Study, a population-based longitudinal study involving 140 centenarians of a total of 186 identified in the study´s geographical area (Ribeiro et al., 2017). This study has a special focus on mental health, successful aging, use of healthcare and social services, and family intergenerational relations. It includes a satellite centenarian study, the PT100 Beira Interior, conducted in the Portuguese countryside region and comprising 101 centenarians (Afonso et al., 2018).

From North Europe, there are the studies of Sweden and Denmark, countries that also present high CRs. In the Swedish Centenarian Study, from the 143 centenarians born 1887–91 and who lived in southern Sweden, 100 agreed to participate (Samuelsson et al., 1997). The purpose of this study was to characterize centenarians in terms of their physical, social, and psychological status, and to predict centenarians' survival (Hagberg & Samuelsson, 2008). Also from Sweden, within the H70 Birth Cohort Studies, there is the H95+ Study, with a representative sample of 95-years-old and older men and women (n = 1020) born in 1901–1911 and living in Gothenburg (Maelstrom Research, 2020). In the third follow-up, started in 2001, the participants were aged 100 years old and a wide variety of data was collected through

Table 4.1 ICC centenarians' studies (non-exhaustive list)

European Countries	Study
Austria	
Belgium	
Bulgaria	
Croatia	
Cyprus	
Czechia	
Denmark	Danish longitudinal centenarian study
Estonia	
Finland	
France	French centenarian study
Germany	Heidelberg centenarian study
Greece	
Hungary	Hungarian centenarian study
Ireland	
Italy	CaT—centenari a trieste Monzino 80+ study
Latvia	
Lithuania	
Luxembourg	
Malta	
Netherlands	100-plus study
Poland	Polish centenarian study
Portugal	PT100—Oporto centenarian study
Romania	
Slovakia	
Slovenia	
Spain	
Sweden	Swedish centenarian study Gothenburg 95+ study
United Kingdom	Cognitive function and aging study

(continued)

cognitive and physical measures, biosamples (e.g., blood) and others. In the Danish Longitudinal Centenarian Study, centenarians from different birth cohorts identified through the Danish Civil Registration System were consecutively studied and compared. The 1895 birth cohort included 276 eligible participants (aged 100 years old) and 207 effectively evaluated; the 1905 cohort included 439 eligible participants (aged 99–100 years old) and 256 effectively evaluated; and, more recently, the 1915 birth cohort, first surveyed in 2010 when they were 95 years old, and second surveyed

Table 4.1 (continued)

Countries outside Europe	Study
USA	Georgia centenarian study Fordham centenarian study Longevity gene project 90+ study (90+) Oregon brain aging study
China	Hong Kong centenarian study
South Korea	Korean centenarian study
Japan	Okinawa centenarian study Tokyo centenarian study
Australia	Sydney centenarian study

Source (Centre for Healthy Brain Ageing, 2020; Iowa State University, 2020) | last update: information not available | reference year: 2020 | download date: 11 May 2020

in 2015, comprised individuals who lived to celebrate their 100th anniversary—498 eligible participants divided in two main areas (West and East cohorts) and 364 effectively evaluated (Rasmussen et al., 2017). Main topics considered for evaluation concern genetics, morbidity, and functional status (Epidemiology Biostatistics and Biodemography Department of Public Health University of Southern Denmark, 2019).

From Central Europe, in Germany, the Heidelberg Centenarian Studies are two population-based investigations conducted 11 years apart, with individuals born around 1900 and around 1911. In the first Heidelberg Centenarian Study (HD100-I), from the 281 eligible persons (age 100 years) a direct contact was established with 156 persons or their relatives. Main addressed domains were cognitive functioning, functional health and care, and subjective well-being. In the second Heidelberg Centenarian Study (HD100-II), following the same sampling procedures as in the first study, of the 485 eligible individuals 112 were interviewed at their residence. The main aim was to examine cohort differences in cognition and physical functioning/care between the two cohorts (Rott, Jopp, & Boerner, 2016). Close to Germany, there is the 100-plus Study Netherlands, an on-going prospective cohort study of Dutch centenarians who self-reported to be cognitively healthy. Three hundred and thirty-two centenarians were evaluated until September 2018, of whom almost 30% (n = 92) agreed to post mortem brain donation. The main focus of this study is cognitive decline, more specifically the protective factors that explain the physiology of long-term preserved cognitive health (Holstege et al., 2018).

Within the set of countries presenting a lower CR there is the Hungarian Centenarian Study, one of the first centenarian studies of the world. Demographic data of 218 centenarians were collected, and 123 were examined by a medical team (Beregi & Klinger, 1989). Presented as having a sociomedical and demographic focus, this centenarian study has information on immunological, clinical laboratory, and morphological profiles, as well as on general data about living conditions at such advanced ages (Regius, 1990). In the Polish Centenarian Study, 346 subjects aged

100 and more years old were visited, and biological material was collected from 285 subjects. The main aims of this study were to assess health status and environmental determinants of extreme ageing, and to collect biological material for examining selected aspects of longevity, including genetic factors (Mossakowska et al., 2008).

The Cognitive Function and Aging Study, from the United Kingdom, is currently listed as a study of the ICC-Dementia. Although including more than 18.000 participants aged over 65 years, since its beginning in the late 1980s subsets of the cohort have been contacted for 1, 2, 6 and 8 year follow ups, and the whole sample was contacted for a 10 year follow up. In this sense, they have a subgroup of participants who are 100 or more years old. The special focus is on dementia and cognitive decline (Cognitive Function & Ageing Society, 2020).

In general, regardless of the conjoint efforts been conducted within the illustrative examples of 5-COOP or ICC, several researchers have defended the idea of a wider network of centenarian studies, the Worldwide Centenarian Consortium (WWC100+), to adequately address the complexity of extreme longevity (Franceschi, Passarino, Mari, & Monti, 2017). Sporadic contributions from researchers interested in exceptional longevity are in fact increasing throughout the world, and bringing together all available contributions has been stated as an important step for addressing several methodological pitfalls that tend to characterize the study of this age group (Sachdev, Levitan, & Crawford, 2012), and for maximizing the generalizability of findings. Either coming from the use of convenience samples that include participants exceptionally old and with the ultimate aim of providing a description of this population and the richness of their long-lived lives (e.g., narrative collection of cases), or by using secondary data from large-scale institutional databases following very specific research interests (e.g., oldest old inpatient profiles in hospitals), the fact is that there is a growing number of studies focusing on the centenarian population (CFA, 2019; Willcox et al., 2010).

4.2   Overview of European Studies on Centenarians

As previously exposed, several efforts have been made to increase the available knowledge on the centenarian population throughout the world and within Europe in particular. Although most published centenarian studies consist of small convenience samples of people being interested and who volunteer to participate in scientific inquiries that explore cases of extreme and "exceptional" longevity, some studies have started to adopt representative or quasi-representative approaches that are expected to provide a more accurate profile on the health and social circumstances of the centenarian population. Also, the use of population-based samples (vs. convenience samples), in being recognized as one of the commonly acknowledged methodological issues of particular importance in the study of the oldest old—along with age verification, use of appropriate control groups, individual differences among centenarians, and moderating and mediating mechanisms (Poon & Perls, 2007), has improved the quality of the available knowledge. Regardless of these methodological and design

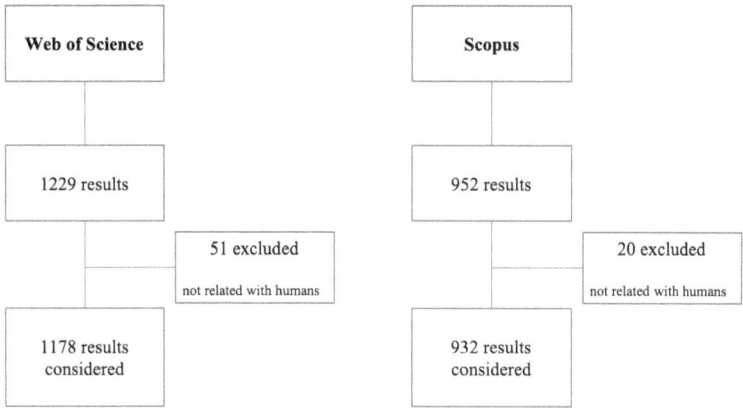

Fig. 4.1   Scope search flowchart

dilemmas in conducting centenarian research, the types and amount of information on centenarians is multidisciplinary and expectedly vast.

In order to provide an overview of the number and main research areas of the centenarian studies conducted in the list of European countries included in this book, we next present findings from a scope search conducted in two academic databases (see methodological note in Chap. 2). Considering the inclusion criteria, since 2000, 1229 and 952 documents were identified in the Web of Science and Scopus, respectively. Excluding results not related with humans (e.g., centenarian anniversaries within Archaeological research, or studies related with Agriculture), 1178 and 932 results were considered (Fig. 4.1). In both sources, the majority of the retrieved documents are classified as articles (76.3% in Web of Science; 72.4% in Scopus). Reviews represent 9.3% of the total results in Web of Science and 11.8% in Scopus. Other types of documents (e.g., meeting abstracts, letters to the editor, editorials, books and book chapters) present percentages lower than 9%.

Figure 4.2 shows the evolution of publications, by database source, in the topic of "centenarians" across EU countries in the last 20 years. In overall, both sources presented the same linear trend, increasing gradually over time. Since this review only covers four months of 2020, the line goes down in this year, as expected. The higher increasements can be observed after 2010, with relative maximum reached at 2013, 2016 and 2019.

The significant increase of the number of publications from 2010 onwards can be possibly linked to the recognition of ageing, and very long lives in particular, as an important issue in the worldwide political agenda. Here we can highlight the United Nation's Sustainable Development Agenda presented in 2015, which established 17 goals aimed to end poverty, protect the planet, and improve the lives and prospects of everyone, everywhere (United Nations, n.d.-b). Within this agenda, population ageing is expected to become one of the most significant social transformations of the twenty-first century, with implications for nearly all sectors of society (United

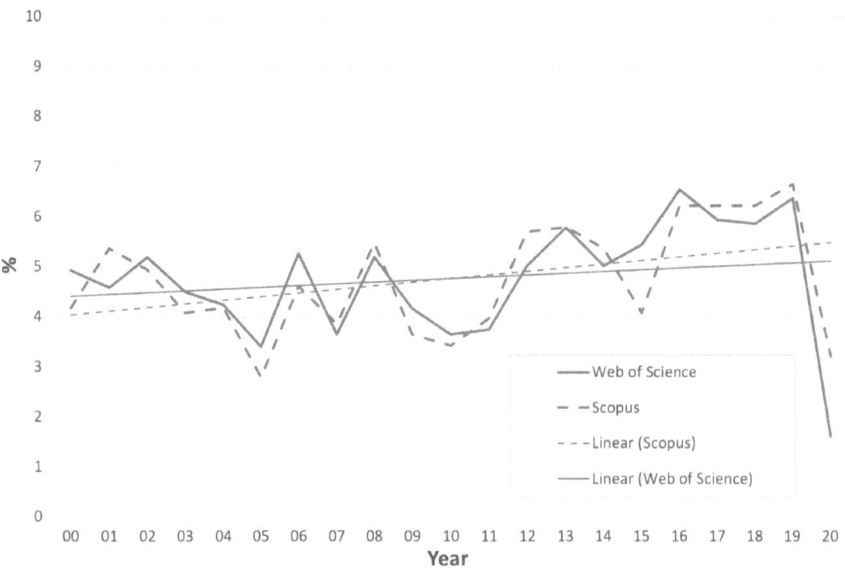

Fig. 4.2 Publications in EU countries in the period 2000–2020 (*Source* Web of Science and Scopus I last update: 8 May 2020 I reference year: 2000–2020 I download date: 8 May 2020)

Nations, n.d.-a). Also the World Health Organization has several contributions aimed to promote healthy ageing. Close to that period, in 2016, the organization presented the Global Strategy and Action Plan on Ageing and Health, which was considered as a significant step forward to achieving the vision that all people can live long and healthy lives (World Health Organization, 2017). At an European level, the European Commission in the Horizon 2020, the biggest EU Research and Innovation programme and financial instrument available between 2014 and 2020 defined "Health, Demographic Change, and Wellbeing" as a European Commission Societal Challenge (European Comission, n.d.) (European Comission, n.d.). In order to respond to this challenge, research and innovation aimed to keep older people active and independent for longer periods, and supporting the development of new, safer and more effective interventions became a priority. The several calls for proposals or actions launched in last years within this challenge were aimed to foster collaboration and coordination in Research and Development (R&D) in Europe. Many financed projects were directed for those people facing some of the challenges of ageing and for those who care for older people if they need help, namely in areas of smart solutions for ageing well and active assisted living. The European Union's Horizon 2020 has contributed to the development of products and services designed to improve quality of life but also to the development of studies to achieve a better understanding of (healthy) ageing, which is fundamental to guide policy recommendations. In both types of projects, many of the outputs were scientific publications on topics related with this challenge (Health, Demographic Change, and Wellbeing) and involving longevity and oldest old participants.

Another aspect that may be associated with the particular increase in the number of publications after 2010 is the availability of data from CENSUS (in 2011). Census have a long tradition in the EU countries. Historically, the development of census by each country was related with several factors, such as information needs or financial cost of the operation. In 2011, for the first time, EU's legislation was defined to achieve high-quality data in EU countries, making the data more comparable. After this year, census allowed comparative studies between EU countries, focusing on population groups about which less information is otherwise available, as it is the case of the centenarian population. The use of this source of information by researchers may have resulted in several indexed publications (e.g., Mašanović et al., 2009; Ribeiro, Teixeira, Araújo, & Paúl, 2016; The Italian Multicentric Study on Centenarians, 1998), but also its public availability may have reinforced the attention given to this population as the increase of their representativeness (and of the oldest old subgroup in general) became more notorious.

When taking a closer look at the number of publications by country (Fig. 4.3), we can observe a noteworthy disparity. Interestingly, the top 10 countries with the highest record of publications are the ones with centenarian studies listed in the International Centenarian Consortium (Table 4.1), with the exception of Spain, that comes in the third place in both Web of Science and Scopus databases. However, despite the current absence of a specific Spanish centenarian study within the ICC, there are several researchers investigating particular issues related with this population in that country. This is the case of the Spanish Centenarian Study Group, from Valencia, that is interested in the molecular mechanisms of longevity (Borras et al., 2016), and of the Cardiac and Clinical Characterization of Centenarians study, with researchers from different cities (Madrid, Mallorca, Cáceres, Lugo, Pontevedra, Guadalajara, and

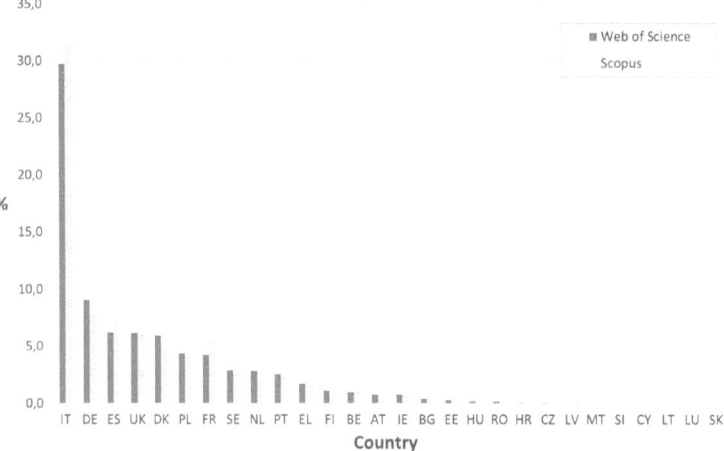

Fig. 4.3 Publications by EU country in the period 2000–2020 (*Source* Web of Science and Scopus l last update: 8 May 2020 l reference year: 2000–2020 l download date: 8 May 2020)

Valladolid) that have looked at the cardiac health of centenarians (Martínez-Sellés, 2015). There is a study from the Universidad Europea de Madrid about the genotype of centenarians from the Spanish central area that included a comparison with Japanese cohorts (Pinós et al., 2014), and studies based on electronic health records that have paid attention to the clinical, drug use and healthcare use characteristics of Spanish centenarians (Gimeno-Miguel et al., 2019). Specific medical topics as the incidence of hip fracture (Barceló et al., 2018) and Venous thromboembolism (Lacruz et al., 2016) have also received the researcher's attention.

Worthwhile underlining is Italy's pole position. This country clearly stands out with a difference of more than 20% in relation to the country occupying the second place in the rank—Germany. The high number of centenarians in Italy may contribute to the existence of more research on this age group. As previously presented, two of the ICC studies are from this country, but there are other important centenarian studies, such as the Italian Multicentric Study on Centenarians (Franceschi et al., 2000), and the AKEA Study (Deiana et al., 1999; Poulain et al., 2004). Also, it is known that the Italian National Institute of Statistics (Istat), in agreement with the International Database on Longevity (IDL), set up an enquiry identified as the Semi-Super-Centenarian (SSC) survey specifically devoted at the validation of all individuals (dead or alive) who had reached or exceeded the age of 105 years on January 1st of the years 2009–2018 (Caselli, Battaglini, & Capacci, 2018; Caselli, Battaglini, & Capacci, 2020). These and other studies, along with the fact that Italy has the Sardinian Longevity Blue Zone—a longevity hotspot that comprises a cluster of 14 villages nestled around the highest mountain of the island, with their epicentre in the village of Villagrande Strisaili (Pes & Poulain, 2016)—may be contributing to the great interest and attention given to centenarians as a research topic among several Italian academics (e.g. Fastame, Hitchcott, & Penna, 2017; Salaris, 2015).

France, although being the country with highest CR (Sect. 3.1), appears as the seventh country with more publications in the last 20 years in the considered databases. The first and unique French Centenarian Study dated 1991 (Allard, Lebre & Robine, 1998; Robine & Allard, 1998), and in addition to the fact that publications coming from this important study are from the end of the 20th century, the majority were written in French. More recently, some independent groups of French researchers presented new knowledge about this population, namely in the areas of genetics, medicine and social sciences. In the area of genetics, some studies had as main purpose the identification of genes or mutations present in French centenarians (Blanché, Cabanne, Sahbatou, & Thomas, 2001; Coppin et al., 2003). In terms of medical sciences, the relation between longevity and higher levels of fibronectin and Lp(a) in plasma were also studied (Labat-Robert et al., 2000; Thillet et al., 1998). On the subject of social sciences, differences between genders in terms of *generational habitus* received attention from researchers of the Inserm, a French public research organization dedicated to human health (Balard, Beluche, Romieu, Willcox, & Robine, 2011).

Some EU countries were found not to have publications in the databases considered for the scope search, or to present a very low number: Malta, Slovenia, Cyprus, Lithuania, Luxembourg and Slovakia. These are the countries with a lower proportion of centenarians, but what also may be justifying such a scarceness of publication is the fact that in many of these countries there is no incentive for publication in indexed journals.

Finally, looking at the affiliation countries of the publications involved in the publications, it was possible to identify several collaborations between EU researchers and other researchers from all around the world. The non-EU countries with higher representativity in such collaborations were USA (e.g. Bae et al., 2018), Switzerland (e.g., Robine & Cubaynes, 2017), Japan (e.g. Pinos et al., 2014), China (e.g. Tuikhar et al., 2019) and Australia (e.g., Ribeiro, Teixeira, Araujo, Afonso, & Pachana, 2015).

As for the research areas being covered by centenarian research in the EU countries, considering those defined by Web of Science, we can observe that the great majority (up to 92.5%) corresponds to Life Sciences & Biomedicine. Only 3.9% are from Social Sciences, 3.1% from Technology, 0.3% from Physical Sciences, and 0.2% from Arts & Humanities. This particular database overrepresents publications for medical sciences and biology, with little focus on psychosocial topics, which may partially explain the obtained results. Still within this database source, expectedly, the "Geriatrics & Gerontology" area represents more than 50% of the obtained results, followed by biological areas, such as "Cell Biology" (13.4%), "Biochemistry and Molecular Biology" (7.0%) and "Genetics Heredity" (6.7%). In terms of medical issues, Cardiology, Endocrinology and Neurology are the specialities with greater numeric expression. Related with the other Research Domains, only Social Sciences was represented in the top-15 with the contributions of Psychology accounting for 3.2% of the publications. As the research areas from this source address very specific topics (such as medical specialities), the category "Other" includes a large number of publications, adding 27.5% of the total results.

In Scopus, the results are divided in four subject areas: Life Sciences (45.3%), Health Sciences (41.3%), Social Sciences (10.1%), and Physical Sciences (3.3%). Following the trend observed in Web of Science, Life Sciences and Health Sciences represent more than 85% of the results, however the Social Sciences here presents a greater contribution, reaching 10% of the total results. The research areas of "Biochemistry, Genetics and Molecular Biology" (60.7%) and "Medicine" (59.2%) appear in the first positions of the top-15. The topic "Aging" is part of the first research area, justifying the expressiveness of this category. The categories used in this source are more comprehensive and generic, resulting in two differentiating aspects of Web of Science: the other domains were represented in the top-15, such as Social Sciences with the areas "Social Sciences" (10.5%), "Arts and Humanities" (3.0%) and "Psychology" (3.0%) and Physical Sciences with the areas "Chemistry" (1.3%), "Environmental Science" (1.3%), "Computer Science" (0.8%) and Chemical Engineering" (0.6%); the category "Other" comprise a small number of results (2.3%).

Table 4.2 Top-15 research areas

Web of science	%	Scopus	%
Geriatrics and gerontology	53.1	Biochemistry, genetics and molecular biology	60.7
Cell biology	13.4	Medicine	59.2
Biochemistry molecular biology	7.0	Nursing	12.2
Genetics heredity	6.7	Social sciences	10.5
Cardiovascular system cardiology	5.1	Neuroscience	6.9
Endocrinology metabolism	5.0	Immunology and microbiology	5.6
Neurosciences neurology	5.0	Agricultural and biological sciences	3.2
Immunology	4.5	Arts and humanities	3.0
Science technology other topics	4.0	Psychology	3.0
General internal medicine	3.6	Pharmacology, toxicology and pharmaceutics	2.7
Psychology	3.2	Multidisciplinary	2.5
Public environmental occupational health	2.5	Chemistry	1.3
Nutrition dietetics	2.3	Environmental science	1.3
Psychiatry	2.2	Computer science	0.8
Physiology	1.9	Chemical engineering	0.6
Other	27.5	Other	2.3

Source Web of Science and Scopus I last update: 8 May 2020 I reference year: 2000–2020 I download date: 8 May 2020

Table 4.2 shows the top-15 research areas from the two sources according to their original categorization of publications.

A note is important to be made on the limitations of this scoped search. Firstly, the time period considered might be excluding results related with studies developed in the 20th century (such as the Hungarian and the French Centenarian Studies). Additionally, the inclusion criteria related with the language (English only) certainly produced skewed search for some countries. For example, France is one country with substantial publications related with centenarians, most of them possibly written and published in French. The use of only two of the most common citation database, that consider only indexed journals, may have excluded other important works related with this topic. Finally, additional key terms could be considered in order to obtain a more exhaustive list of all possible results related with research in centenarians (e.g. oldest-old).

Despite these limitations, in general, we could observe that the growing interest in extreme longevity is leading to the production of a good amount of scientific literature that covers distinct subjects within centenarian research. Worthwhile noting is that along with demography and genetics, researchers have mainly pointed their attention towards the cognitive, social, and physical condition of the oldest-old, more than the psychological health (Scelzo, 2019). Also, there is a dearth of literature examining the housing and financial circumstances of centenarians (Serra et al., 2011), as well as on their care provision, (unmet) needs, and complexities that often involve long-term informal caregiving and increasing demands of social and health services. These constitute undoubtedly important topics to be considered in an international research agenda for centenarian studies in the years to come (Jopp, Boerner, Ribeiro, & Rott, 2016).

Chapter 5
Final Remarks

The information we have presented throughout this book provided elementary but important information on the centenarian population in most EU countries. In general, we could observe that EU centenarians are mostly women and widowed, tend to live in urban areas and in the community. Greater heterogeneity was found in what regards education and household profiles. Circulatory system diseases were identified as the main cause of death for near centenarians (95+). Despite this overall summarization, the fact that this book has used open databases presents some limitations. First, the information provided by each database cannot be crossed out, which constrained the analysis to be eminently descriptive. Additionally, the information for people aged 100 years or older is very restricted. For example, mental health and physical health information, though of great interest, are not available in this kind of databases. Notwithstanding these limitations, we believe on the relevance of profiling the centenarian population at a European level as it provides a general picture of this growing and challenging segment of the population.

From the overview presented, we certainly assume that much knowledge is needed to fully capture centenarians' current condition and deeply understand the course of such long lives. Within this context, centenarian studies currently being conducted in most EU countries as well as from abroad (e.g., USA, Japan, Australia) constitute excellent opportunities to look for answers about human nature, possible trajectories of health and well-being, genetic and self-determinism in life. The available information coming from such studies will certainly provide insightful information on how the European society, both at a broader and country specific levels, can optimize the quality of life of the growing number of long-lived individuals and their families.

Although some resemblances between EU countries may exist in terms of longevity as we could observe in the findings presented throughout this book, an overall distinctive "EU centenarian profile" most certainly does not exist. Variations are potentially due not only to country characteristics that range from political constellations, socioeconomic determinants, and health services access, to name a

L. Teixeira et al., *Centenarians*, SpringerBriefs in Aging,
https://doi.org/10.1007/978-3-030-52090-8_5

few factors, but also to eminently personal characteristics like genetics and psychological resilience that strongly account for the so-called heterogeneity of the oldest old population.

Countries are now focusing on emergent and critical policy questions in an attempt to respond to the needs and challenges of the longevity revolution (ILC-BR, 2015), including social and health services, assisted living and caregiving, but also social protection, pensions and financial investments that are crucial for long-term investments and policy implementations. Throughout the EU, centenarians will be a challenging issue in both the developed and the emerging economies.

As societies are beginning to adjust to the projected future of larger numbers of oldest old individuals and smaller numbers of younger ones who have historically comprised the support base, new forms of social protection for older people are needed. Moreover, the trend in the increasing number of centenarians has important consequences for the way each one of us perceives longevity, but also for the way in which societal infrastructures like education, workplaces, housing, transport, and health and social care will need to be adapted to their needs. As Lesson (2018) recently mentioned when reflecting on the drivers and implications of living to 100 years and beyond, our perception of old age might as well need a dramatic reappraisal (Lesson, 2018).

For the future, it will be important to develop synergies between public and private institutions (governments, universities, hospitals) around the specific topic of exceptional longevity in order to deepen the knowledge about this population. Given the differences in methodological options (e.g., inclusion criteria, identification of potential participants, information collected) that characterize centenarian studies around the world and within Europe, it is often very difficult to compare the richness of data already available. Methodologically standardizing studies within the European context, for instance, would be an important step forward in the establishment of valid international comparisons. Collecting large and thorough heterogeneous samples as it is the intention of research groups like the ICC or the 5-COOP seems to be a road to take, as it will make possible to provide reliable answers to several epidemiological questions that small-scale studies have not the possibility to offer.

Along with facing the challenges of exceptional longevity from a socio-political and economic perspective, also individuals must prepare themselves for the challenges of long lives and to the intergenerational dynamics associated with this phenomenon. Within the scope of personal resources influencing exceptional longevity it is somehow expected that there will be an increase in psychological mortality in the very old that is characterized by loss of intentionality, identity, psychological autonomy and control (Baltes & Smith, 2003). Although the human body has not been made to last, even at age 100 psychological factors such as beliefs and attitudes are still good predictors of well-being, underlining the notion that psychological mechanisms are an important potential for successful coping with limitations and adversities at advanced ages.

Accumulating negative conditions represent a serious challenge to adaptability and are likely to reduce the quality of life; therefore increasing the available knowledge on core concepts like loneliness and social isolation that are intrinsic to human losses once one gets to very old ages, and on the psychological strengths one may have for positive adaptation to adversities are key aspects for further study within exceptional longevity research.

With the actual general knowledge raised along this book and the findings already made available on the geographic areas where unusually high proportions of long-lived individuals have been observed—the Blue Zones (Ogliastra in Sardinia, Okinawa in Japan, the Nicoya peninsula in Costa Rica, and the island of Ikaria in Greece) (Poulain et al., 2013), among other studies, we may finish emphasizing that along with adequate and integrated social and health policies and raising living conditions, social cohesion and community ties largely contribute to an ageing (European) society where centenarians are expected to be more numerically expressive, but, most importantly, are expected to have the opportunity to continue living meaningful lives.

References

Afonso, R. M., Ribeiro, O., Vaz Patto, M., Loureiro, M., Loureiro, M. J., Castelo-Branco, M., & Rocha, C. (2018). Reaching 100 in the countryside: Health profile and living circumstances of portuguese centenarians from the beira interior region. *Current Gerontology and Geriatrics Research, 2018.* https://doi.org/10.1155/2018/8450468.

Allard, M., Lebre, V., & Robine, J.-M. (1998). Jeanne Calment, from Van Gogh's time to ours, 122 extraordinary years. WH Freeman and Company.

Allard, M., & Robine, J.-M. (2000). *Les centenaires français. Etude de la Fondation IPSEN. 1990–2000. Rapport final*: Serdi Edition.

Andersen-Ranberg, K., & Jeune, B. (2006). The Danish longitudinal centenarian study. In Z. Yi, E. M. Crimmins, Y. Carrière, J.-M. Robine (Eds.), *Longer life and healthy aging* (pp. 151–172). Springer.

Arosio, B., Ostan, R., Mari, D., Damanti, S., Ronchetti, F., Arcudi, S., Scurti, M., Franceschi, C., & Monti, D. (2017). Cognitive status in the oldest old and centenarians: A condition crucial for quality of life methodologically difficult to assess. Mechanisms of Ageing Development, 165(Part B), 185–194. https://doi.org/10.1016/j.mad.2017.02.010.

Bae, H., Gurinovich, A., Malovini, A., Atzmon, G., Andersen, S. L., Villa, F., Barzilai, N., Puca, A., Perls, T. T., & Sebastini, P. (2018). Effects of FOXO3 polymorphisms on survival to extreme longevity in four centenarian studies. *Journals of Gerontology Series A-Biological Sciences and Medical Sciences, 73*(11), 1439–1447. https://doi.org/10.1093/gerona/glx124.

Balard, F., Beluche, I., Romieu, I., Willcox, D. C., & Robine, J.-M. (2011). Are men aging as oaks and women as reeds? A behavioral hypothesis to explain the gender paradox of French centenarians. *Journal of Aging Research, 2011.* https://doi.org/10.4061/2011/371039.

Ballarino, G., Meschi, E., & Scervini, F. (2013). The expansion of education in Europe in the 20th century. *AIAS, GINI Discussion Paper 83.*

Baltes, P. B., & Smith, J. (2003). New frontiers in the future of aging: From successful aging of the young to the dilemmas of the fourth age. *Gerontology, 49*(2), 123–135. https://doi.org/10.1159/000067946.

Barceló, M., Francia, E., Romero, C., Ruiz, D., Casademont, J., & Torres, O. H. (2018). Hip fractures in the oldest old. Comparative study of centenarians and nonagenarians and mortality risk factors. *Injury, 49*(12), 2198–2202. https://doi.org/10.1016/j.injury.2018.09.043.

Barford, A., & Dorling, D. (2006). Life expectancy: Women now on top everywhere. *BMJ, 332*(7545), 808. https://doi.org/10.1136/bmj.332.7545.808.

Barthold, D., Nandi, A., Mendoza Rodríguez, J. M., & Heymann, J. (2014). Analyzing whether countries are equally efficient at improving longevity for men and women. *American Journal of Public Health, 104*(11), 2163–2169. https://doi.org/10.2105/AJPH.2013.301494.

Beregi, E., & Klinger, A. (1989). Health and living conditions of centenarians in Hungary. *International Psychogeriatrics, 1*(2), 195–200. https://doi.org/10.1017/S1041610289000207.

Berzlanovich, A. M., Keil, W., Waldhoer, T., Sim, E., Fasching, P., & Fazeny-Dörner, B. (2005). Do centenarians die healthy? An autopsy study. *The Journals of Gerontology: Series A, 60*(7), 862–865. https://doi.org/10.1093/gerona/60.7.862.

Blanché, H., Cabanne, L., Sahbatou, M., & Thomas, G. (2001). A study of French centenarians: Are ACE and APOE associated with longevity? *Comptes Rendus de l'Académie des Sciences-Series III-Sciences de la Vie, 324*(2), 129–135. https://doi.org/10.1016/S0764-4469(00)01274-9.

Boerner, K., Jopp, D., Park, M.-K., & Rott, C. (2016). Whom do centenarians rely on for support? Findings from the second Heidelberg centenarian study. *Journal of Aging & Social Policy, 28*(3), 165–186. https://doi.org/10.1080/08959420.2016.1160708.

Bohacek, R., Crespo, L., Mira, P., & Pijoan-Mas, J. (2015). The educational gradient in life expectancy in Europe: Preliminary evidence from SHARE. In A. Börsch-Supan, T. Kneip, H. Litwin, M. Myck, & G. Weber (Eds.), *Ageing in Europe-supporting policies for an inclusive society*. Gruyter.

Borras, C., Abdelaziz, K. M., Gambini, J., Serna, E., Inglés, M., de la Fuente, M., Garcia, I., Matheu, A., Sanchís, P., Belenguer, A., Errigo, A., Avellana, J.-A., Barettino, A., Lloret-fernández, C., Flames, N., Pes, G., Rodriguez-Mañas, L., & Viña, J. (2016). Human exceptional longevity: Transcriptome from centenarians is distinct from septuagenarians and reveals a role of Bcl-xL in successful aging. *Aging (Albany NY), 8*(12), 3185–3208. https://doi.org/10.18632/aging.101078.

Bozzini, S., & Falcone, C. (2017). Genetic factors associated with longevity in humans. In J. Dorszewska & W. Kozubski (Eds.), *Senescence-physiology or pathology*. IntechOpen.

Brandão, D., Ribeiro, O., Afonso, R. M., & Pául, C. (2019). Regional differences in morbidity profiles and health care use in the oldest old: Findings from two Centenarian studies in Portugal. *Archives of Gerontology and Geriatrics, 82,* 139–146. https://doi.org/10.1016/j.archger.2019.02.009.

Brodaty, H., Woolf, C., Andersen, S., Barzilai, N., Brayne, C., Cheung, K. S.-L., Corrada, M. M., Crawford, J. D., Daly, C., Gondo, Y., Hagberg, B., Hirose, N., Holstege, H., Kawas, C., Kaye, J., Kochan, N. A., Lau, B., H.-P., Lucca, U., Marcon, G., . . ., Sachdev, P. S. (2016). ICC-dementia (International Centenarian Consortium-dementia): An international consortium to determine the prevalence and incidence of dementia in centenarians across diverse ethnoracial and sociocultural groups. *BMC Neurology, 16*(1), 16–52. http://doi.org/10.1186/s12883-016-0569-4.

Bumpass, L., Sweet, J., & Martin, T. C. (1990). Changing patterns of remarriage. Journal of Marriage and the Family, 52(3), 747–756.

Caselli, G., Battaglini, M., & Capacci, G. (2020). Beyond one hundred: A cohort analysis of italian centenarians and semi-supercentenarians. *The Journals of Gerontology. Series B, Psychological Sciences and Social Sciences, 75*(3), 591–600. http://doi.org/10.1093/geronb/gby033.

Caselli, G., Battaglini, M., & Capacci, G. (2018). Cohort analysis of gender gap after one hundred years old: The role of differential migration and survival trajectories. *Journal of Ageing Science, 6*(3), 1–7. https://doi.org/10.35248/2329-8847.19.07.199.

Caselli, G., & Lipsi, R. (2016). Survival differences among the oldest old in Sardinia: Who, what, where, and why? *Demographic Research, 14,* 267–294. https://doi.org/10.4054/DemRes.2006.14.13.

Centraal Bureau voor de Statistiek. (2020). Annual population estimates as of January 1st, ages 100+ (100, 108+), 1950–2000. Data received via e-mail on 1 November 2001 from Meer van der Lubbe at the CBS Netherlands. (Data obtained through the Human Mortality Database, www.mortality.org or www.humanmortality.de, on 2 January 2020).

Central Statistical Bureau (CSB) of Latvia. (2020). Official population estimates 1971–2004. (Data obtained through the Human Mortality Database, www.mortality.org or www.humanmortality.de, on 2 January 2020).

Centre for Healthy Brain Ageing. (2020). About ICC-Dementia. Retrieved from https://cheba.unsw.edu.au/consortia/icc-dementia.

CFA. (2019). The Centre for Policy on Ageing's selected readings–Centenarians. Retrieved from http://www.cpa.org.uk/information/readings/centenarians.pdf.

Chen, Y.-C., Hu, H.-Y., Fan, H.-Y., Kao, W.-S., Chen, H.-Y., & Huang, S.-J. (2019). Where and how centenarians die? The role of hospice care. *American Journal of Hospice and Palliative Medicine, 36*(12), 1068–1075. https://doi.org/10.1177/1049909119845884.

Clayton, G. M., Dudley, W. N., Patterson, W. D., Lawhorn, L. A., Poon, L. W., Johnson, M. A., & Martin, P. (1994). The influence of Rural/Urban residence on health in the oldest-old. *The International Journal of Aging & Human Development, 38*(1), 65–89. https://doi.org/10.2190/XV9T-JKJK-HBTA-62H1.

Cognitive Function & Ageing Society. (2020). Welcome to CFAS. Retrieved from http://www.cfas.ac.uk.

Coppin, H., Bensaid, M., Fruchon, S., Borot, N., Blanche, H., & Roth, M. (2003). Longevity and carrying the C282Y mutation for haemochromatosis on the HFE gene: Case control study of 492 French centenarians. *BMJ, 327*(7407), 132–133. https://doi.org/10.1136/bmj.327.7407.132.

Danmarks Statistik. (2020). (Data obtained through the Human Mortality Database, www.mortality.org or www.humanmortality.de, on 2 January 2020).

de Lorgeril, M., Salen, P., Paillard, F., Laporte, F., Boucher, F., & de Leiris, J. (2002). Mediterranean diet and the French paradox: Two distinct biogeographic concepts for one consolidated scientific theory on the role of nutrition in coronary heart disease. *Cardiovascular Research, 54*(3), 503–515. https://doi.org/10.1016/s0008-6363(01)00545-4.

Deiana, L., Ferrucci, L., Pes, G. M., Carru, C., Delitala, G., Ganau, A., Mariotti, S., Mieddu, A., Pettinato, S., Putzu, P., Franceschi, C., & Baggio, G. (1999). AKEntAnnos. The Sardinia study of extreme longevity. *Aging Clinical and Experimental Research, 11*(3), 142–149.

Dupraz, J., Andersen-Ranberg, K., Fors, S., Herr, M., Herrmann, F. R., Wakui, T., Jeune, B., Robine, J.-M., Saito, Y., & Santos-Eggimann, B. (2020). Use of healthcare services and assistive devices among centenarians: Results of the cross-sectional, international 5-COOP study. *BMJ Open, 10*(3). http://dx.doi.org/10.1136/bmjopen-2019-034296.

Epidemiology Biostatistics and Biodemography Department of Public Health University of Southern Denmark. (2019). List of publications: Danish 100 year-olds. Retrieved from https://www.sdu.dk/en/om_sdu/institutter_centre/ist_sundhedstjenesteforsk/forskning/epidemiologi/forskningsprojekter/centenarians/publ_cent.

European Comission. (2008). *Poverty And Social Exclusion In Rural Areas.* Retrieved from https://ec.europa.eu/social/BlobServlet?docId=2087&langId=en.

European Comission. (2014). *The Census Hub: Easy and flexible access to European census data.* Retrieved from https://ec.europa.eu/eurostat/documents/4031688/6285607/KS-02-14-480-EN-N.pdf/05b4ca91-1f72-4dbb-ae2c-d3a07f56d795.

European Comission. (n.d.). What is Horizon 2020? Retrieved from https://ec.europa.eu/programmes/horizon2020/what-horizon-2020.

European Union. (2019). *Ageing Europe. Looking at the lives of older people in the EU.* Retrieved from https://ec.europa.eu/eurostat/documents/3217494/10166544/KS-02-19%E2%80%91681-EN-N.pdf/c701972f-6b4e-b432-57d2-91898ca94893.

Eurostat. (2018). *International Standard Classification of Education (ISCED).* Retrieved from https://ec.europa.eu/eurostat/statistics-explained/index.php/International_Standard_Classification_of_Education_(ISCED).

Eurostat. (2019). *Mortality and life expectancy statistics.* Retrieved from https://ec.europa.eu/eurostat/statistics-explained/pdfscache/1274.pdf.

Falkingham, J., Evandrou, M., McGowan, T., Bell, D., & Bowes, A. (2010). Demographic Issues, Projections and Trends: Older people with high support needs in the UK. *Joseph Rowntree Foundation.*

Fastame, M. C., Hitchcott, P. K., & Penna, M.-P. (2017). Does social desirability influence psychological well-being: Perceived physical health and religiosity of Italian elders? A developmental approach. *Aging & Mental Health, 21*(4), 348–353. https://doi.org/10.1080/13607863.2015.1074162.

Franceschi, C., Motta, L., Valensin, S., Rapisarda, R., Franzone, A., Berardelli, M., Motta, M., Monti, D., Bonafé, M., Ferruci, L., Deiana, L., Pes, G. M., Carru, C., Desole, M. S., Barbi, C.,

Sartoni, G., Gemelli, C., Lescai, F., Olivieri, F., . . ., Baggio, G. (2000). Do men and women follow different trajectories to reach extreme longevity? *Aging Clinical and Experimental Research, 12*(2), 77–84. https://doi.org/10.1007/BF03339894.

Franceschi, C., Passarino, G., Mari, D., & Monti, D. (2017). Centenarians as a 21st century healthy aging model: A legacy of humanity and the need for a world-wide consortium (WWC100+). *Mech Ageing Dev, 165*(Pt B), 55–58. https://doi.org/10.1016/j.mad.2017.06.002.

Fredman, L., & Lyons, J. (2012). Measurement of social factors in aging research. In A. Newman & J. Cauley (Eds.), *The epidemiology of aging* (pp. 135–151). Springer.

Gavrilov, L. A., & Gavrilova, N. S. (2000). Validation of exceptional longevity. *Population and Development Review, 26*(2), 403–404.

Gavrilov, L. A., & Gavrilova, N. S. (2019). Late-life mortality is underestimated because of data errors. *PLoS Biology, 17*(2). https://doi.org/10.1371/journal.pbio.3000148.

Gerontology Research Group. (2020). Validated Living Supercentenarians. Retrieved from http://www.grg.org/SC/SCindex.html.

Gimeno-Miguel, A., Clerencia-Sierra, M., Ioakeim, I., Poblador-Plou, B., Aza-Pascual-Salcedo, M., González-Rubio, F., Herrero, R. R., & Prados-Torres, A. (2019). Health of Spanish centenarians: A cross-sectional study based on electronic health records. *BMC Geriatrics, 19*, 226. https://doi.org/10.1186/s12877-019-1235-7.

Gómez-Redondo, R., & González, J. M. G. (2010). Emergence and verification of supercentenarians in Spain. In H. Maier, J. Gampe, B. Jeune, J.-M. Robine, & J. W. Vaupel (Eds.), *Supercentenarians* (pp. 151–171). Springer.

Hagberg, B., & Samuelsson, G. (2008). Survival after 100 years of age: A multivariate model of exceptional survival in Swedish centenarians. *The Journals of Gerontology Series A: Biological Sciences and Medical Sciences, 63*(11), 1219–1226. https://doi.org/10.1093/gerona/63.11.1219.

Hank, K. (2007). Proximity and contacts between older parents and their children: A European comparison. *Journal of Marriage and Family, 69*(1), 157–173. https://doi.org/10.1111/j.1741-3737.2006.00351.x.

Herm, A., Cheung, K., & Poulain, M. (2012). Emergence of oldest old and centenarians: Demographic analysis. *Asian Journal of Gerontology and Geriatrics, 7*(1), 19–25.

Herr, M., Jeune, B., Fors, S., Andersen-Ranberg, K., Ankri, J., Arai, Y., Cubaynes, S., Santos-Eggimann, B., Zekry, D., Parker, M., Saito, Y., Hermann, F., Robine, J.-M., & 5-COOP group (2018). Frailty and associated factos among centenarians in the 5-COOP countries. *Gerontology, 64*(6), 521–531. https://doi.org/10.1159/000489955.

Herrmann, F. R., & Zekry, D. (2017). Comparison of centenarians' characteristics among 5 countries, the Oldest Old Project (5-COOP). *Innovations in Aging, 1*(suppl_1), 1347–1348. https://doi.org/10.1093/geroni/igx004.4948.

Holstege, H., Beker, N., Dijkstra, T., Pieterse, K., Wemmenhove, E., Schouten, K., Thiessens, L., Horsten, D., Rechtuijt, S., Sikkes, S., A. van Poppel, F. W., Meijers-Heijboer, H., Hulsman, M., & Scheltens, P. (2018). The 100-plus Study of cognitively healthy centenarians: Rationale, design and cohort description. *European Journal of Epidemiology, 33*(12), 1229–1249. https://doi.org/10.1007/s10654-018-0451-3.

Human Mortality Database. (2020). Retrieved from www.mortality.org or www.humanmortality.de.

ILC-BR. (2015). *Active Ageing: A Policy Framework in Response to the Longevity Revolution.* Retrieved from http://ilcbrazil.org/portugues/wp-content/uploads/sites/4/2015/12/Active-Ageing-A-Policy-Framework-ILC-Brazil_web.pdf.

International Database on Longevity. (2020, 09/12/2019). Retrieved from https://www.supercentenarians.org.

Iowa State University. (2020). International Centenarian Consortium. Retrieved from https://research.hs.iastate.edu/international-centenarian-consortium/.

Jeune, B., Robine, J. M., Young, R., Desjardins, B., Skytthe, A., & Vaupel, J. W. (2010). Jeanne Calment and her successors. Biographical notes on the longest living humans. In H. Maier, J. Gampe, B. Jeune, J.-M. Robine, & J. W. Vaupel (Eds.), *Supercentenarians* (pp. 285–323). Springer.

Jong-Gierveld, J., Valk, H., & Blommesteijn, M. (2011). Living Arrangements Of Older Persons And Family Support In More Developed Countries. *Population Bulletin of the United Nations, 43/44*(special issue), 193–217.

Jopp, D., Boerner, K., Ribeiro, O., & Rott, C. (2016). Life at age 100: An international research agenda for Centenarian studies. *Journal of Aging & Social Policy, 28*(3), 133–147. https://doi.org/10.1080/08959420.2016.1161693.

Jopp, D., Boerner, K., & Rott, C. (2016). Health and disease at age 100—findings from the second Heidelberg Centenarian study. *Deutsches Aerzteblatt International, 113*, 203–210. https://doi.org/10.3238/arztebl.2016.0203.

Jopp, D., & Rott, C. (2006). Adaptation in very old age: Exploring the role of resources, beliefs and attitudes for centenarians' happiness. *Psychology and Aging, 21*(2), 266–280. https://doi.org/10.1037/0882-7974.21.2.266.

Labat-Robert, J., Marques, M. A., N'Doye, S., Alpérovitch, A., Moulias, R., Allard, M., & Robert, L. (2000). Plasma fibronectin in French centenarians. *Archives of Gerontology and Geriatrics, 31*(2), 95–105. https://doi.org/10.1016/S0167-4943(00)00071-6.

Lacruz, B., Tiberio, G., Núñez, M. J., López-Jiménez, L., Riera-Mestre, A., Tiraferri, E., Verhamme, P., Mazzolai, L., González, J., Monreal, M., & RIETE Investigators (2016). Venous thromboembolism in centenarians: Findings from the RIETE registry. *European Journal of Internal Medicine, 36*, 62–66. https://doi.org/10.1016/j.ejim.2016.07.025.

Lesson, G. W. (2018). Living to 100 years and beyond: Drivers and implications. *European View, 17*(1), 44–51. https://doi.org/10.1177/1781685818758965.

Lucca, U., Tettamanti, M., Tiraboschi, P., Logroscino, G., Landi, C., Sacco, L., Garrì, M., Ammesso, S., Biotti, A., Gargantono, E., Piedicorcia, A., Mandelli, S., Riva, E., Galbussera, A. A., & Recchia, A. (2020). Incidence of dementia in the oldest-old and its relationship with age: The Monzino 80-plus population-based study. *Alzheimer's & Dementia, 16*(3), 472–481. https://doi.org/10.1016/j.jalz.2019.09.083.

Lundström, H. (2020). Annual population register counts as of December 31st, 1970–1989. Unpublished computer files. (Data obtained through the Human Mortality Database, www.mortality.org or www.humanmortality.de, on 2 January 2020).

Lyberaki, A., Tinios, P., Mimis, A., & Georgiadis, T. (2013). Mapping population aging in Europe: How are similar needs in different countries met by different family structures? *Journal of Maps, 9*(1), 4–9. https://doi.org/10.1080/17445647.2012.752334.

MacDonald, M. (2007). Social support for Centenarians' health, psychological well-being, and longevity. *Annual Review of Gerontology and Geriatrics, 27*(1), 107–127.

Maelstrom Research. (2020). The Gothenburg H70 Birth Cohort Studies. Retrieved from https://www.maelstrom-research.org/mica/individual-study/h-70#.

Maier, H., Gampe, J., Jeune, B., Robine, J. M., & Vaupel, J. W. (2010). *Supercentenarians*. Springer.

Martin, P., & Martin, M. (2002). Proximal and distal influences on development: The model of developmental adaptation. *Developmental Review, 22*(1), 78–96. https://doi.org/10.1006/drev.2001.0538.

Martinez-Gonzalez, M., & Martín-Calvo, N. (2016). Mediterranean diet and life expectancy; beyond olive oil, fruits and vegetables. *Current Opinion in Clinical Nutrition & Metabolic Care, 19*(6), 401–407. https://doi.gov/10.1097/MCO.0000000000000316.

Martínez-Sellés, M., García de la Villa, B., Cruz-Jentoft, A. J., Vidán, M. T., Gil, P., Cornide, L., Cortés, M. R., González Guerrero, J. L., Barros Cerviño, S. M., Castro, O. D., Pareja, T., Sánchez, E., Cimera, D., Vigara, M., Balaguer, J., Mogollón Jiménez, M. V., Fernández, F. V., Juanatey, C. G., Fernández, A. T., . . ., Diaz, J. L. (2015). Centenarians and their hearts: A prospective registry with comprehensive geriatric assessment, electrocardiogram, echocardiography, and follow-up. *American Heart Journal, 169*(6), 798–805. https://doi.org/10.1016/j.ahj.2015.03.005.

Mašanović, M., Šogorić, S., Kolčić, I., Curić, I., Smoljanović, A., Ramić, S., Cala, M., & Polasek, O. (2009). The geographic patterns of the exceptional longevity in Croatia. *Collegium Antropologicum, 33*(1), 147–152.

Medford, A., Christensen, K., Skytthe, A., & Vaupel, J. W. (2019). A cohort comparison of lifespan after age 100 in Denmark and Sweden: Are only the oldest getting older? *Demography, 56*(2), 665–677.

Meslé, F., & Vallin, J. (2002). Comment améliorer la précision des tables de mortalité aux grands âges? Le cas de la France. *Population, 57*(4), 601–629.

Meslé, F., & Vallin, J. (2002). Mortality in Europe: The divergence between east and west. *Population, 57*(1), 157–197.

Mestheneos, E. (2011). Ageing in place in the European union. *IFA Global Ageing, 7*(2), 17–23.

Mirowsky, J. (2003). *Education, social status, and health.* Routledge.

Modig, K., Andersson, T., Vaupel, T., Rau, R., & Ahlbom, A. (2017). How long do centenarians survive? Life expectancy and maximum lifespan. *Journal of Internal Medicine, 282*(2), 156–163. doi:10.1111/joim.12627.

Mossakowska, M., Barcikowska, M., Broczek, K., Grodzicki, T., Klich-Raczka, A., Kupisz-Urbanska, M., Podsiadly-Moczydlowska, T., Sikora, E., Szybinska, A., Wieczorowska-Tobis, K., Zyckowska, J., & Kuznicki, J. (2008). Polish Centenarians Programme–multidisciplinary studies of successful ageing: Aims, methods, and preliminary results. *Experimental Gerontology, 43*(3), 238–244. https://doi.org/10.1016/j.exger.2007.10.014.

National Statistical Institute. (2020). Annual population by age (single-year age group) and sex. Unpublished tables. (Data obtained through the Human Mortality Database, www.mortality.org or www.humanmortality.de, on 2 January 2020).

Neumayer, E., & Plumper, T. (2016). Inequalities of income and inequalities of longevity: A cross-country study. *American Journal of Public Health, 106*(1), 160–165. doi: 10.2105/AJPH.2015.302849.

OECD. (2011). *Health Reform: Meeting the Challenge of Ageing and Multiple Morbidities.* Retrieved from https://www.oecd.org/health/health-systems/health-reform-9789264122314-en.htm.

OECD. (2017). *Health at a Glance 2017: OECD Indicators.* Retrieved from https://www.oecd-ilibrary.org/docserver/health_glance-2017-en.pdf?expires=1595001377&id=id&accname=guest&checksum=22E92036F802D94733B4AAB0E55C0842.

Olshansky, S. J., Antonucci, T., Berkman, L., Binstock, R. H., Boersch-Supan, A., Cacioppo, J. T., Carnes, B. A., Carstensen, L. L., Fried, L. P., Goldman, D. P., Jackson, J., Kohli, M., Rother, J., Zheng, Y., & Rowe, J. (2012). Differences in life expectancy due to race and educational differences are widening, and many may not catch up. *Health Affairs, 31*(8), 1803–1813. https://doi.org/10.1377/hlthaff.2011.0746.

Ostan, R., Monti, D., Gueresi, P., Bussolotto, M., Franceschi, C., & Baggio, G. (2016). Gender, aging and longevity in humans: An update of an intriguing/neglected scenario paving the way to a gender-specific medicine. *Clinical Science, 130*(19), 1711–1725. doi: 10.1042/CS20160004.

Pes, G., & Poulain, M. (2016). Blue Zones. In N. Pachana (Ed.), *Encyclopedia of geropsychology.* Springer.

Pignolo, R. J. (2019). Exceptional human longevity. *Mayo Clinic Proceedings, 94*(1), 110–124. https://doi.org/10.1016/j.mayocp.2018.10.005.

Pinós, T., Fuku, N., Cámara, Y., Arai, Y., Abe, Y., Rodríguez-Romo, G., Garatachea, N., Santos-Lozano, A., Miro-Casas, E., Ruiz-Meana, M., Otagui, I., Murakami, H., Miyachi, M., Garcia-Dorado, D., Hinohara, K., Andreu, A. L., Kimura, A., Hirose, N., & Lucia, A. Ruiz-Meana, M. (2014). The rs1333049 polymorphism on locus 9p21. 3 and extreme longevity in Spanish and Japanese cohorts. *Age, 36*(2), 933–943. http://doi.org/10.1007/s11357-013-9593-0.

Poon, L. W., & Cheung, K. (2012). Centenarian research in the past two decades. *Asian Journal of Gerontology and Geriatrics, 7*(1), 8–13.

Poon, L. W., Clayton, G. M., Martin, P., Johnson, M. A., Courtenay, B. C., Sweaney, A. L., Merriam, B. B., Pless, B. S., & Thielman, S. B. (1992). The Georgia centenarian study. *The International Journal of Aging and Human Development, 34*(1), 1–17. doi: 10.2190/8M7H-CJL7-6K5T-UMFV.

Poon, L. W., Hagberg, B., & Martin, P. (2004). A Brief Chronology of the International Centenarian Consortium. Retrieved from https://www.gerontology.iastate.edu/wp-content/uploads/2015/04/ICShistory-to-2014.pdf.

Poon, L. W., Martin, P., Bishop, A., Cho, J., da Rosa, G., Deshpande, N., Hensley, R., MacDonald, M., Margrett, J., Randall, G. K., Woodard, J. L., & Miller, L. S. (2010). Understanding centenarians' psychosocial dynamics and their contributions to health and quality of life. *Current Gerontology and Geriatrics Research, 2010*. https://doi.org/10.1155/2010/680657.

Poon, L. W., & Perls, T. (2007). The trials and tribulations of studying the oldest old. *Annual Review of Gerontology and Geriatrics, 27*(1), 1–10.

Poulain, M. (2010). On the age validation of supercentenarians. In H. Maier, J. Gampe, B. Jeune, J. M. Robine, & J. W. Vaupel (Eds.), *Supercentenarians* (pp. 3–30). Springer.

Poulain, M., Herm, A., & Pes, G. (2013). The Blue Zones: Areas of exceptional longevity around the world. *Vienna Yearbook of Population Research, 11*, 87–108.

Poulain, M., Pes, G., Grasland, C., Carru, C., Ferruci, L., Baggio, G., et al. (2004). Identification of a geographic area characterized by extreme longevity in the Sardinia island: The AKEA study. *Experimental Gerontology, 39*(9), 1423–1429. https://doi.org/10.1016/j.exger.2004.06.016.

Ramirez, F., & Boli, J. (1987). The political construction of mass schooling: European origins and worldwide institutionalization. *Sociology of Education, 60*(1), 2–17.

Randall, G. K., Martin, P., McDonald, M., Poon, L. W., Georgia Centenarian Study, Jazwinski, S. M., Green, R. C., Gearing, M., Markesbery, W. R., Woodard, J. L., Johnson, M. A., Tenover, J. S., Siegler, I. C., Rodgers, W. L., Hausman, D. B., Rott, C., Davey, A., & Arnold, J. (2010). Social resources and longevity: Findings from the Georgia centenarian study. *Gerontology, 56*(1), 106–111. https://doi.org/10.1159/000272026.

Rasmussen, S. H., & Andersen-Ranberg, K. (2016). Health in Centenarians. In N. Pachana (Ed.), *Encyclopedia of geropsychology*. Springer, Singapore.

Rasmussen, S. H., Andersen-Ranberg, K., Thinggaard, M., Jeune, B., Skytthe, A., Christiansen, L., Vaupel, J. W., McGue, M., & Christensen, K. (2017). Cohort Profile: The 1895, 1905, 1910 and 1915 Danish Birth Cohort Studies-secular trends in the health and functioning of the very old. *International Journal of Epidemiology, 46*(6), 1746–1746j. https://doi.org/10.1093/ije/dyx053.

Rechel, B., Roberts, B., Richardson, E., Shishkin, S., Shkolnikov, V. M., Leon, D.A., Bobak, M., Karanikolos, M., & McKee, M. (2013). Health and health systems in the commonwealth of independent states. *The Lancet, 381*(9872), 1145–1155. https://doi.org/10.1016/S0140-6736(12)620 84-4.

Regius, O. (1990). Study of living conditions of centenarians. In E. Beregi (Eds.), *Centenarians in hungary* (pp. 151–161). Karger Publishers.

Ribeiro, O., Araújo, L., Teixeira, L., Brandão, D., Duarte, N., & Paúl, C. (2017). PT100-Oporto Centenarian Study. In N. Pachana (Ed.), *Encyclopedia of geropsychology*. Springer.

Ribeiro, O., Araújo, L., Teixeira, L., Duarte, N., Brandão, D., Martin, I., & Paúl, C. (2016). Health status, living arrangements and service use at 100. Findings from the Oporto Centenarian Study. *Journal of Aging & Social Policy, 28*(3), 148–164. https://doi.org/10.1080/08959420.2016.116 5582.

Ribeiro, O., Brandão, D., Araújo, L., Teixeira, L., Paúl, C., & Poulain, M. (2020). Exceptional siblings: The andrade brothers. *Journal of the American Geriatrics Society, 68*(5), 1112–1114. https://doi.org/10.1111/jgs.16396.

Ribeiro, O., Teixeira, L., Araujo, L., Afonso, R. M., & Pachana, N. (2015). Predictors of anxiety in centenarians: Health, economic factors, and loneliness. *International Psychogeriatrics, 27*(7), 1167–1176. https://doi.org/10.1017/s1041610214001628.

Ribeiro, O., Teixeira, L., Araújo, L., & Paúl, C. (2016). Health profile of centenarians in Portugal: A census-based approach. *Population Health Metrics, 14*(1). http://doi.org/10.1186/s12963-016-0083-3.

Robine, J.-M, Romieu, I., & Allard, M. (2003). [French centenarians and their functional health status]. *Presse medicale, 32*(8), 360–364.

Robine, J.-M., & Cubaynes, S. (2017). Worldwide demography of centenarians. *Mechanisms of Ageing and Development, 165*(Part B), 59–67. https://doi.org/10.1016/j.mad.2017.03.004.

Robine, J.-M. (2019). Successful aging and the longevity revolution. In R. Fernández-Ballesteros, A. Benetos, & J. M. Robine (Eds.), *The Cambridge handbook of successful aging* (pp. 27–38). Cambridge University Press.

Robine, J.-M., & Allard, M. (1998). The oldest human. *Science, 279*(5358), 1831. https://doi.org/10.1126/science.279.5358.1831h.

Robine, J.-M., & Allard, M. (1999). Jeanne Calment: Validation of the duration of her life. In B. Jeune & J. W. Vaupel (Eds.), *Validation of exceptional longevity*. Odense University Press.

Robine, J.-M., Allard, M., Herrmann, F. R., & Jeune, B. (2019). The real facts supporting Jeanne Calment as the oldest ever human. *The Journals of Gerontology: Series A, 74*(Supplement_1), S13–S20. https://doi.org/10.1093/gerona/glz198.

Robine, J.-M., & Caselli, G. (2005). An unprecedented increase in the number of centenarians. *Genus, 61*(1), 57–82.

Robine, J.-M., Cheung, S. L. K., Saito, Y., Jeune, B., Parker, M. G., & Herrmann, F. R. (2010). Centenarians today: New insights on selection from the 5-COOP study. *Current Gerontology and Geriatrics Research, 2010*. https://doi.org/10.1155/2010/120354.

Robine, J.-M., & Vaupel, J. W. (2002). Emergence of supercentenarians in low-mortality countries. *North American Actuarial Journal, 6*(3), 54–63. https://doi.org/10.1080/10920277.2002.10596057.

Rott, C., Jopp, D., & Boerner, K. (2016). Heidelberg centenarian studies. In N. Pachana (Ed.), *Encyclopedia of geropsychology*. Springer.

Sachdev, P., Levitan, C., & Crawford, J. (2012). Methodological issues in centenarian research: Pitfalls and challenges. *The Asian Journal of Gerontology and Geriatrics, 7*(1), 44–48.

Salaris, L. (2015). Differential mortality in a long-living community in Sardinia (Italy): A cohort analysis. *Journal of Biosocial Science, 47*(4), 521–535. https://doi.org/10.1017/S0021932014000224.

Samuelsson, S.-M., Alfredson, B. B., Hagberg, B., Samuelsson, G., Nordbeck, B., Brun, A., Gustafson, L., & Risberg, J. (1997). The Swedish Centenarian Study: A multidisciplinary study of five consecutive cohorts at the age of 100. *The International Journal of Aging and Human Development, 45*(3), 223–253. 10.2190/XKG9-YP7Y-QJTK-BGPG

Santana, P. (2015). *A Geografia da Saúde da População. Evolução nos últimos 20 anos em Portugal Continental*. Retrieved from https://www.uc.pt/fluc/gigs/GeoHealthS/Geografia-saude-populacao-20anos-Portugal.pdf.

Scelzo, A. (2019). The complex journey into longevity: Helping the oldest old to live happier. *International Psychogeriatrics, 31*(11), 1527–1529. https://doi.org/10.1017/S1041610219000929.

Serra, V., Watson, J., Sinclair, D., & Kneale, D. (2011). *Living Beyond 100: A report on centenarians*. Retrieved from https://ilcuk.org.uk/wp-content/uploads/2019/03/Living-Beyond-100-A-report-on-centenarians.pdf.

Statistical Office of the Slovak Republic. (2020). Population estimates (January 1st) by age and sex. Data received in Excel format. (Data obtained through the Human Mortality Database, www.mortality.org or www.humanmortality.de, on 2 January 2020).

Statistics Finland. (2020). Population estimates for years 1941–1995 (originally 31 Dec 1940–1994) have been received in the form of a computer file directly from Statistics Finland. They are supposedly compiled from annual publications on Finnish population structure: Väestörakenne YYYY (Population Structure YYYY), Suomen virallinen tilasto (OfficialStatistics of Finland), Väestö (Population). YYYY+1:6. Helsinki: Tilastokeskus (Statistics Finland); YYYY+1. (Data obtained through the Human Mortality Database, www.mortality.org or www.humanmortality.de, on 2 January 2020).

Statistik Austria. (2020). Annual population estimates by age and sex. (Data obtained through the Human Mortality Database, www.mortality.org or www.humanmortality.de, on 2 January 2020).

Tettamanti, M., & Marcon, G. (2018). Cohort profile:'Centenari a Trieste'(CaT), a study of the health status of centenarians in a small defined area of Italy. *BMJ Open, 8*(2). http://dx.doi.org/10.1136/bmjopen-2017-019250.

The Italian Multicentric Study on Centenarians. (1998). Laboratory parameters of Italian centenarians. *Archives of Gerontology and Geriatrics, 27*(1), 67–74. https://doi.org/10.1016/S0167-4943(98)00100-9.

Thillet, J., Doucet, C., Chapman, J., Herbeth, B., Cohen, D., & Faure-Delanef, L. (1998). Elevated lipoprotein (a) levels and small apo (a) isoforms are compatible with longevity: Evidence from a large population of French centenarians. *Atherosclerosis, 136*(2), 389–394. https://doi.org/10.1016/S0021-9150(97)00217-7.

Tuikhar, N., Keisam, S., Labala, R. K., Imrat, Ramakrishnan, P., Arunkumar, M. C., Ahmed, G., Biagi, E., & Jeyaram, K. (2019). Comparative analysis of the gut microbiota in centenarians and young adults shows a common signature across genotypically non-related populations. *Mechanisms of Ageing and Development, 179*, 23–35. http://doi.org/10.1016/j.mad.2019.02.001.

United Nations. (2015). *World Population Ageing.* Retrieved from https://www.un.org/en/development/desa/population/publications/pdf/ageing/WPA2015_Report.pdf.

United Nations. (2017). *Principles and Recommendations for Population and Housing Censuses. Revision 3.* Retrieved from https://unstats.un.org/unsd/demographic-social/Standards-and-Methods/files/Principles_and_Recommendations/Population-and-Housing-Censuses/Series_M67rev3-E.pdf.

United Nations. (2019a). *2018 Demographic Yearbook.* Retrieved from https://unstats.un.org/unsd/demographic-social/products/dyb/dybsets/2018.pdf.

United Nations. (2019b). *World Population Prospects 2019: Online Edition.*

United Nations. (n.d.-a). Ageing. Retrieved from https://www.un.org/en/sections/issues-depth/ageing/index.html.

United Nations. (n.d.-b). The Sustainable Development Agenda. Retrieved from https://www.un.org/sustainabledevelopment/development-agenda/.

Vacante, M., D'Agata, V., Motta, M., Malaguarnera, G., Biondi, A., Basile, F., Malaguarnera, M., Gagliano, C., Drago, F., & Salamone, S (2012). Centenarians and supercentenarians: A black swan. Emerging social, medical and surgical problems. BMC Surgery, 12(Suppl 1). doi: 10.1186/1471-2482-12-S1-S36.

Vallin, J., & Meslé, F. (2001). Tableau I-C-1: Population par sexe et âge (de 0 à 100 ans), au 1 janvier, de 1899 à 1998, avec deux estimations selon le territoire pour les années de changement de territoire [revised post-publication]. In: Tables de mortalité françaises pour les XIXe et XXe siècles et projections pour le XXIe siècle. Paris: Instit national d'études démographiques. (Data obtained through the Human Mortality Database, www.mortality.org or www.humanmortality.de, on 2 January 2020).

Vaupel, J. W. (2010). Biodemography of human ageing. *Nature, 464*, 536–542.

Vaupel, J. W., & Jeune, B. (1995). The emergence and proliferation of centenarians. In J. W. Vaupel & B. Jeune (Eds.), *Exceptional Longevity: From Prehistory to the Present* (pp. 1–13). Odense University Press.

Willcox, D. C., Willcox, B. J., Hsueh, W.-C., & Suzuki, M. (2006). Genetic determinants of exceptional human longevity: Insights from the Okinawa Centenarian Study. *Age, 28*(4), 313–332. doi: 10.1007/s11357-006-9020-x.

Willcox, D. C., Willcox, B. J., & Poon, L. W. (2010). Centenarian studies: Important contributors to our understanding of the aging process and longevity. *Current Gerontology and Geriatrics Research, 2010*, 1–6. https://doi.org/10.1155/2010/484529.

Willcox, D. C., Willcox, B. J., Wang, N.-C., He, Q., Rosenbaum, M., & Suzuki, M. (2008). Life at the extreme limit: Phenotypic characteristics of supercentenarians in Okinawa. *The Journals of Gerontology Series A: Biological Sciences and Medical Sciences, 63*(11), 1201–1208. https://doi.org/10.1093/gerona/63.11.1201.

Wilmoth, J., Skytthee, A., Friou, D., & Jeune, B. (1996). The oldest man ever? A case study of exceptional longevity. *The Gerontologist, 36*(6), 783–788.

World Health Organization. (2011). *ICD-10: International statistical classification of diseases and related health problems: 10th Revision. World Health Organization.*

World Health Organization. (2017). Global strategy and action plan on ageing and health. Retrieved from https://www.who.int/ageing/WHO-GSAP-2017.pdf?ua=1.

World Health Organization. (2020). Retrieved from https://www.who.int/healthinfo/mortality_dat a/en/.

Young, R., & Kroczek, W. (2019). Validated living worldwide supercentenarians 112+, Living and recently deceased: February 2019. *Rejuvenation Research, 22*(1), 79–81. doi: 10.1089/rej.2019.2179.

Yu, R., Tam, W., & Woo, J. (2017). Trend of centenarian deaths in Hong Kong between 2001 and 2010. *Geriatrics and Gerontology International, 17*(6), 931–936. doi: 10.1111/ggi.12812.

Zak, N. (2019). Evidence that jeanne calment died in 1934—not 1997. *Rejuvenation Research, 22*(1), 3–12. doi: 10.1089/rej.2018.2167.